Jeffery Sm... ...eer since 1970. After... ...nd on tall blocks of flat... ...ects of houses, in pa... ...ficient houses for the... ...d for the National Buil... ...ernment closed this in 1983. For 30... ...he ran a consultancy in east London advising on and designing structures including repair of defective houses and, with architects, enhancing Victorian houses with an extra room since the kids were growing up. He was also a Party Wall Surveyor. In the early 1970s he was a one of a group that set up Open Christmas shelters for the homeless in London. He lived in a Quaker inspired community in east London for 17 years.

Best wishes
Jeffery Smith

INSULATE!!

Mass retrofit of houses in England and Wales done by clubs of owner-occupiers

How to fully insulate your home along with your neighbours, and not just the lofts!

Jeffery Smith

Published independently by MR Editions

© Copyright Jeffery Smith 2023 … but further thinking and discussion welcome

ISBN 978 1 3999 6386 2

Networking via: mr4insulate@gmail.com

Acknowledgements:
- Hilary, my wife
- All the east London builders I met.

...from the websites, July 2023

- The World Meteorological Organization said the planet experienced the hottest few days on record in the first few days of July, after a June that was the hottest on record, according to the European Space Agency.

- A study recently published in 'Nature Medicine' said more than 60,000 people died because of last year's summer heatwaves across Europe, with the highest mortality rates seen in Italy, Spain and Portugal.

- Temperatures in Rome, which is packed with tourists, are poised to climb to 42C or 43C on Tuesday. Sixteen Italian cities, including Rome, Florence, Bologna, Bari, Cagliari and Palermo, have been put on "red alert" by the health ministry, meaning the heat is so intense it poses a threat to the health of the entire population. Night-time temperatures remain above 20C, making it a struggle for people to sleep.

- With a Saturday temperature of 37.4C, Fort Good Hope in the Northwest Territories saw "the hottest temperature recorded that far north in Canada," said an Environment Canada meteorologist.

- Parts of Europe, Asia and North America are preparing for heat on Monday that threatens to break records, drive wildfires and has prompted health warnings and evacuations.

- Europe could record its hottest-ever temperature this week on Italy's islands of Sicily and Sardinia where a high of 48C (118F) is predicted, source: the European Space Agency.

- The US National Weather Service warned a "widespread and oppressive" heatwave in southern and western states was expected to peak.

- Japan has issued heatstroke alerts affecting tens of millions of people, as near-record high temperatures hit several parts of the country, with other areas pummelled by torrential rain.

- Global sea surface temperatures (SST) reached a new record. The global SST of 20.98C (69.76F) is a record 0.638C hotter than the 1991s mean.

INSULATE!!

Mass retrofit of houses in England and Wales done by clubs of owner-occupiers

How to fully insulate your home along with your neighbours, and not just the lofts!

There is no true creation without the unforeseeable becoming necessity.
PIERRE BOULEZ

CONTENTS

Preface 13

Introduction 17

Chapter One 21
- things necessary to know before you and your neighbours decide to set up a retrofit club.
- Scenes: Anxieties, a WhatsApp group, a general meeting of the street, progress, the enquiry consultant or broker, money, initial arrangements;
- CASE STUDY The first mass retrofit, Green Street.

Chapter Two 33
– possibilities for the retrofit club set-up
- Scenes: structural arrangement, the retrofit club, funding mass retrofit, insurers, retrofit builders, life plan approach, surveying and assessments, houses for the 22nd century, government role, approvals, search and remediate, condensation, full assurance, insurance risks, nerves of steel, other professionals, building wise – need for remediation;
- CASE STUDY: The second mass retrofit, Thunberg St.

Chapter Three 51

– including a picture of the sorts of things your experts need to do to set up contracts to carry out mass retrofit.
- Scenes: surveys, features, not to cover over defects, current insulation levels, external wall cladding, a new type of gutter, fixing of cladding, appearance, features of an external wall, ground storeys, other aspects of external walls, internal wall lining, roofs and their insulation, how a loose timbered roof was built, loose timber roofs in use, altering a roof, central valley roofs, trussed rafter roofs, solar panels, chimneys, incorrect removal of chimney breasts, loft insulation, a water tank, loft storage, roof safeguards, retrofitting of ground floors;
- CASE STUDY: The 200th mass retrofit, Carson Street.

Chapter Four: 87

- *The Event* - about what will be happening during the key time that retrofit work is actually underway in your street.
- Scenes: the event, advance preparations, initial arrival of the builder, full arrival of the builder, climate emergency, external wall retrofit methods, umbrella for your building, internal wall cladding, a week of retrofit, draughts and windows, gutters, inside the house, non-male operatives, the garden, the big departure;
- CASE STUDY: The Case Of 20 Dioxide Drive.

Chapter Five: 101
-about finishing off and follow up and how to encourage others.
- Scenes: Scenes: left over houses, tidy up, follow-up, snagging, odds and ends, cavities filled, energy use monitored, passport to the future, how long retrofit will last; time for a party; caution and endeavour.
- CASE STUDY: The future of *Planet Earth*

APPENDICES 109

BIBLIOGRAPHY 157

GLOSSARY 159

ILLUSTRATIONS 161

INDEX 165

PREFACE

Planet Earth is in grave risk of a climate catastrophe. We have all learnt this in the past few years.

You are troubled. You can't help being immensely saddened. You feel desperately *if only I and my family could do something*. You wonder what you can do about your house to mitigate its carbon footprint.

This book says how you can insulate houses, perhaps like yours, better, and not just the loft! You club together with neighbours, bring in expertise and eventually a retrofit builder who takes charge of your street. They install thermal cladding externally above your ground storey window heads - a bit like putting an overcoat over your house. And during working day visits, they fit thinner internal wall insulation to ground storey walls on the inside. They anticipate the need to slightly alter kitchen units, boilers, pipes, cables etc so the insulation can fit in.

They do other necessary things to make the house function correctly, now that it has better thermal insulation. The set-up will allow for human foibles and, for instance, sudden illness of residents.

Costs may be met by a retrofit building society loan paid back over 10 to 20 years by amounts that are similar to recent energy bills.

Insurers are involved to cover unexpected situations, and also to provide workmanship warranties. This is important as full retrofit should not cover over significant cracking or defects due to poor previous alterations. The builder must repair these first.

This way you very significantly reduce emissions harmful to the planet.

Other ways of achieving something like full insulation are mentioned. Some ways to best do this are yet to be devised.

This book

In this book we set out ways of your club doing this in your street and the next one. We outline the big issues and suggest ways and means. We do not know all the answers, but we seek to outline the known issues, and suggest approaches to the unknown. Every case will be different, though often like some other instance.

Following mass retrofit, houses in your street will not be 'net-zero 'but they should be three quarters of the way there! And the realistic hope is that by addition of solar panels (if necessary strengthening roof structures) or heat pumps these houses will become net-zero in carbon production. However such installations are not a subject covered in this book. Also not covered is your boiler, except that after retrofit one with lower heat output should suffice and it will mainly relate to providing domestic hot water. And maybe it will be fuelled by hydrogen.

If such retrofit is widespread the nation can greatly mitigate its damage to the planet caused by the human need for shelter. And, may be, other countries might learn from our example with their housing stock. (We are learning from them.)

There are around 22 million dwellings in England & Wales, most of them houses. It is likely that 80% of these will still exist in 2050. So, retrofitting them nearer to net-zero standards is the only way England and Wales can cut home emissions. Ways of building in Scotland are often different, for instance there may be more single storey houses and more people live in apartments. But some of the principles here will apply to a proportion of Scottish houses.

The retrofit industry will for a time become bigger than the house construction industry. What this book describes for owner occupied houses can happen largely independently of government, but government could pass useful legislation and taxation benefits to ease and encourage things. In particular it can decide that VAT should not be chargeable on retrofit or anything necessary for retrofit, for instance any rectification of defects called for in advance of retrofit.

The Event

The Event is when the retrofit contractor takes over your street for a period of time (the 'contract period') and, on days planned in advance, comes into parts of your house. Others nearby will probably have already had the retro-fit event, others are may be yet to have it. To have it will be a *Big Event* in your lives and that of all involved. Efforts will be in place to minimise unsafe activity. The *Event* might seem a nuisance, but looked at in another way it will be when you and your family start to be nicer to planet Earth on which we live. Carb and Di and their kids quite liked it, though it was something of a bother at times. Read about them and the case of 20 Dioxide Drive at the end of Chapter Four.

INTRODUCTION

Low Carbon Work So Far: New houses

People are experimenting with building new houses whose carbon emissions due to building and in use do not harm the planet. Having very effective insulation is just part of this. People are looking to ways the very act of house construction can emit the lowest output of carbon and greenhouse gasses to Earth's atmosphere.

One of the most developed methods is being developed in Germany and called *Passivhaus*. Practitioners in other countries are adopting this and similar approaches, adapting to circumstances in their country as appropriate. Traditionally, or at least since the mid 19th century, manufacture of building materials to build a house has used large quantities of fossil fuels. For instance, much energy is needed to fire bricks in something like a furnace. And through the twentieth century, cement, or more particularly Portland cement, is used with crushed stone and sand to make concrete. Nowadays almost all foundations are built in concrete. It is used for superstructure beams, columns and other components so that, for instance, larger room sizes are possible.

Last but not least among high energy building products is steel. Ores or recycled steel are heated in furnaces till molten metal is formed. On cooling this can be rolled into plate, railway lines, I section beams, and for other building products. And, of course, steel is used to make domestic products such as washing machines, cars, trucks and many other things we take for granted.

Steel can of course be recycled. And it has been for a century or more, being thrown into a furnace, melted down and eventually rolled out as new. This does though call for a great input of energy. So few advocates of recycling would suggest steel as the best example of recycling!

Existing houses

But what about existing houses? These too often are draughty and poorly insulated. You probably live in one like this, or did once. This book is about how to insulate the houses of England and Wales more fully, so not just the lofts. A significant part of this will be getting rid of draughts. Ultimately this may mean mechanical ventilation with heat recovery (MVHR). With a market of over 20 million homes, MVHR companies may hope to bring out mini units that can be fitted in each room, perhaps below the window. In these outgoing air is blown through a metal drum which takes most heat out of the air to be emitted. Incoming air passes over the same drum and is warmed up by it.

In this context, people are looking at how to convert various forms of existing buildings including houses to have zero-carbon impact or very low zero carbon implications. One way of doing this is called *EnerPHit*.

It will not be possible to convert the mass of British housing to match *Passivhaus* construction. Neither is it feasible to take them all down and rebuild them in the Passivhaus way! Where will people live in the meantime? And isn't a *Passivhaus* more bulky due to having thicker walls, so they won't fit into our tight urban street plans?

Way forward

Fifteen years ago the government asked a Bank of England governor about the UK housing market and why it did not obey the laws of supply and demand like other commodities. Part of her explanation was that on the then present trends, replacing all UK houses would take a thousand years!

In this book we look at a way that can get 20 million houses energy efficient, and this by 2040. This follows a pilot phase up till 2028 which includes a start to learning good practice and pooling experience.

It is important to say that this book is not a treatise on mass-retrofit. Others will write books of that sort based on what is learned from projects, possibly like those in your street.

This book does not tell you every detail. Specialists who the club brings in will do that, taking suitable account of the house types and arrangements they find. Indeed, some techniques mentioned may still need to be perfected. This book may seem repetitive at times. This is so that each section can better stand up for itself!

No more condensation and mould

A spin-off benefit of improved insulation will be that condensation and mould in houses no longer occurs. This in itself will be a significant achievement. Recently a coroner blamed mould in a house for causing the death of a child.

CHAPTER ONE
-things necessary to know before you and your neighbours
set up a retrofit club

Scenes: Anxieties; a WhatsApp group; a general meeting of the street; retro-fit of houses; getting things together; the managing consultant or broker; what a retrofit project consists of; money; cost of borrowing; insurance and permissions; insulate, so far.

Anxieties

You, your family and your neighbours are anxious that not enough is being done to improve houses and counter climate change effects and thereby safeguard Planet Earth[1]. You and some neighbours have discussed things and all agreed in their anxieties. You have heard of similar concern at the other end of your street. More than one WhatsApp group is discussing things. Whilst some talk of 'doing it together', as many are not sure how to go on.

This book may help you to go on. It is based on the writer's several decades experience as a structural engineer and building surveyor much of it in housing often with architects. Sometimes this was with large organizations, some more at

---

[1] This feeling is in line that the House of Commons Committee on Climate Change when it published a report in February 2019 on the future of UK housing. It stated that the UK's housing was "not fit for the future", and that progress had stalled on ensuring it was fit for meeting the challenges of climate change and people's wellbeing.

grass roots level. This is based on experience at his former east London practice and in civil engineering when younger. Some of it was in retrofit called for due to other reasons. For example, 1960 new town terraces built with flat roofs that leaked, so needing pitched roofs to be added, this, together with other improvements, done with residents present in the 1980s.

A general meeting of the street

A general meeting of the street and nearby areas is called. It sets up an action group and agrees club arrangements.

Retrofit of houses

Often many houses in a street are of similar design and built in the same decade. Inner city housing, especially in London, is often 150 years old already and many houses now lived in were lived in before World War 2, some before World War 1. (There is more about houses and their types in the Appendix.)

This book is about houses, not bungalows, not flats. Yes, they need de-carbonizing but in different ways. And yes, this is being done. Tall blocks of flats can be clad and indeed have been for over ten years now. But they must be done in the right way. The dreadful fire at Grenfell Tower, west London, in 2017 is an example of what can go wrong. Seventy-two people died.

Most houses already have loft insulation, meaning that thermal insulating material, usually mineral wool, has been placed over the top floor ceilings. In some cases, a further layer has been placed over this and over the wooden ceiling joists. In all cases it is hoped there remains suitable ventilation lest moisture levels build up in vulnerable parts of roof timbers (dew-points) and timber decay sets in. Needing a new roof is not a clever situation to get into, particularly when one is working to save the planet!

Many recognize that wall insulation comes next. But how, where (internal or external) and with what? Some recognize that even if that is decided there could be locations and zones in a house needing particular consideration and possibly special treatment. Otherwise 'cold bridges[2]' may remain. These might be at the eaves, at corners or at features such as bay windows.

This book suggests that working on a whole street basis should make all these improvements possible to a reduced budget. And compliant with the Building Regulations!

Getting things together

So it is decided to look for professional advice. To float itself the club invites individual house owners to join, at an initial fee, say £100. Enough join. So the club chooses a core group, a committee, to progress things.

Some people will say but we were just about to alter our house to get another room. Then when our kids grow up they can have one each. The club may let them join with this in view, though may have had to devise a strict policy on this sort of thing from the start. But this might be an extra £100 fee to cover admin! Clearly costs of designing and altering must be met by them. In some cases a main retrofit contractor may

---

[2] A cold bridge is a place where the cold gets in noticeably, really where heat leaks out severely. See Glossary.

have a network of builders who can do particular 'one-off' improvements.

Whilst a surveyor at the estate agents has heard of *Passivhaus* and an architect's practice has someone who went on a course and someone else phoned a medium size building company, further progress on tackling the street hits a hard place.

Probably something of a multi-disciplinary help is needed from specialists who could tackle at least say 80 houses as part of a single contract.

So, your retrofit club puts some professional expertise together to search more systematically. But the owners can help form a view on one important aspect: are there sufficient adjoining houses? The answer is probably that most houses will have to have an adjoining house that is part of the scheme. And only a few houses in a terrace can be 'afforded' that do not have a house both sides in the scheme. Unless of course they are end of terrace when it will help greatly if the neighbour's house is in the scheme. As well as economies of scale a major issue will be cold bridges where an adjacent house is not clad sufficiently. In the case of a hundred year old ailing neighbour, it may be possible to fit external wall cladding up to their window reveal, at no cost to them. And a 'flying squad' of over-cladding specialists deals with that house later.

Semi-detached houses should almost always done in pairs. This is so much the case that an owner who is not an occupier might be sought out with a view to inviting them into the project.

The enquiry consultant or broker

Because of complexities anticipated, especially in the early days of mass retrofit, a versatile practice is found that has a number of principals[3] and ideally these are from more than one construction or other relevant discipline. Study of their CVs shows many have experience in larger scale projects and in both the private and public sectors. They are prepared to draw up a systematic method and an enquiry document that can be put before others.

Therefore, they are briefed. They look for medium to large size builders who have either been around some time or seem versatile. Probably it is best if the eventual contract is of a type by the Joint Contract Tribunal (JCT) though once mass retrofit is widespread JCT may do well to write a special form of contract for retrofit.

However, they also look into some unusual aspects which accompany what is after all an unusual situation. Not least is that the work is to be done with residents remaining in occupation.

There is recognition that at least some work on lofts and roofs could mean a temporary roof over is required (an 'umbrella'). Indeed, this will probably be called for so that external walls can dry out. The intention is that most work internally can be done on a room-by-day basis, that is in 9 hour occupations starting around 8am.

What a retrofit project consists of

It is soon recognized that these requirements will determine the nature of the project – it will not be like a normal building project. So, we call it a retrofit project, the word retrofit implying two things: it is done almost as an afterthought but in a calculated way to upgrade what is there; and to a degree it

---

[3] Then if one is on holiday or ill another partner is drafted in.

will be part of everyone's everyday life for a year or more. All personal and family arrangements will have to allow for this, though we do mention exceptions.

Everyday life of course consists of many things, and let us say things not done every day but equally necessary or likely to occur: so there is being awake, being asleep, waking up, taking meals, sex, having a baby, getting kids off to school, getting off to work, going on holiday, caring for granny, being ill, dying, etc. Yes, all this will or could happen in the same street as mass retrofit.

So, the enquiry consultant or broker advises the club to employ people to help keep everyday life running as well as building trade people active: Household Liaison Officers! (HeLOs). Soon the list might be as long as the 'cast list' we include, tongue in cheek, in an appendix.

Money

Last but certainly not least, the retrofit club and the broker look at the money needed to do retrofit. Why not some sort of mortgage using future savings on energy costs to bring about the means for funding it? Not so very different to borrowing from a building society to buy your house. Indeed, many will be doing just that – and retrofit too!

But when you buy a house that is already built if anything goes wrong the building society takes over and sells the house, so their risks are limited.

The costs of retrofitting though will be set against the building, not owners. The Land Registry should be informed. When the house sells the next owners continue with the payment plan. Obviously, you wonder what the cost of retrofit will be. This book cannot say other than it might be between £20,000 and £50,000, exclusive of Value Added Tax. Later we make a strong point that VAT should not be chargeable on retrofit type contracts as saving the planet takes precedence.

Cost of borrowing

Clearly recently increased interest rates could inhibit mass retrofit. The hope is that streets exist in which sufficient adjoining house owners can provide say half the funding, thereby needing to borrow less. The hope will be that sufficient numbers of these can form the kernel for a mass retrofit movement and build up experience. Perhaps family members will lend to or on behalf of house-owning daughters and sons at special or even no rate of interest. Then on the basis that interest rates will reduce a broader front of mass retrofit can 'take off'.

That said, this book points to a way which 'ordinary' owner-occupiers in England and Wales can take a crucial initiative in limiting their carbon footprint without resorting to a government initiative. True, government could help in its taxation policies especially regarding no VAT. It will however be up to others to devise and improve ways for houses that are not occupied by an owner to be de-carbonized. Indeed, to a degree this is already happening. As we have said this book is not about retrofitting flats and apartments.

Insurance and permissions

But retrofit risks are greater so need to be factored in from the start. This means insurance, and clever insurance at that! So there will be an inter-relationship between owners, your club. Let's call the club the "Carbon Street Association of Owners and Residents", the Coal-Abolished Retrofit Building Society, the AllStreet Insurance Society, and (indirectly) Planet Earth Central Re-insurance.

And that's just the players needed to do the work. Ancillary will be: Household Liaison Officers (HeLOs), Heave-it furniture removers, MumsForDeCarb, DadsDefeated, GrannyGrid, GrandadsPuzzled, DeCarb&Cancer, Defiant-90-year-olds, Trans-whatever, I-want-a-quiet-life-anon, young people's clubs according to age. And more.

And local authorities will need to be satisfied, or adjust. Can a fire engine get in? Is a retrofit street more at risk of fire (building sites tend to be)? If a street is closed to make a builder's compound does everyone know the new entry route? Will the temporary car pound where your car must go be handy? Or must there be a mini-bus service? Can waste disposal proceed?

Not least, can the retrofit builder get in? For instance, behind terraces of inner London Victorian terraced houses. The answer could be that as well as looking to use a void house for offices a builder might even advocate removal of a house so that they can get to rear elevations. The plot could be kept as a public garden. Or at the end it could be sold and a *Passivhaus* built!

Insulation, so far

So, when am I going to say something about how to retrofit? The answer is soon. Except that the professionals your retrofit club employs are the ones to decide in any particular circumstance. But, Sterne[4] like, you may notice I instead tell you the story of house insulation so far, or most of the story.

Starting at the top, the roof, the most exposed part of a house.

And it is the top of the top floor ceiling where all the retrofit action has been seen so far. For over 5 decades people have been laying thermal quilting such as mineral wool between ceiling joists. Probably it was 5cm or 7.5cm thick, the ceiling joists probably being 12cm deep, so still standing up visibly. In recent decades thermal insulation mineral quilting thickness has equated to ceiling joist depths, so possibly 12cm thick and there is sometimes another layer over across the ceiling joists.

This makes it hazardous to go into the loft. A good scheme of thermal insulation will have included wooden gangways between the entrance hatch and the water tank. These would be fixed on spacers above ceiling joists. And possibly some lengths to put stored items on, provided the structural engineer consents.

More importantly a good installer will have ensured there is through-ventilation. This ensures no build-up of humidity that could result in the roof timbers decaying.

That's pretty much the story of house insulation so far. Just occasionally a house owner will have added insulation to walls. But was this done in a suitable way? Read on.

---

[4] Laurence Sterne (1713-1768), see Bibliography

CASE STUDY ONE

The First Mass Retrofit

The residents of Green Street formed a committee. Its Chair was Kate who had recently retired as CEO of a housing association. Deputy was Quentin, still a partner in an architect's practice. Stan, a builder, was thought a key person to be part of things. Xenie was treasurer, though initially all that meant was banking subscriptions - £100 per house. It was thought important to have Marie who lived in the Brown Street cul-de-sac which was joining in. And Gertrude the hair-dresser – she seemed likely to hear the sentiments of all involved, or at least all women involved. There were ten houses on Brown Street actually twelve but two were buy-to-rent. With 40 houses in owner occupation on Green Street, 50 was thought a suitable number to set up viable contracts.

Following a very affirmative general meeting of the owners the first job was to appoint managing agents. At least that's what they called them since there were no retrofit consultants in existence yet. They looked at various ways, at medium size architects based in the city for instance. They looked at a firm of chartered surveyors in the city 40 miles away, a practice which was strong on building surveyors, those trained more in building technology than property values. They looked at a set up run by a friend of a vociferous woman at the far end of Brown Street.

An important consideration was that they recognised it may be best to choose another consultant to design and manage things once funding was in place, but a case might be made for the same managing agent to be designers and contract administrators.

In the end they chose Solemn Building Surveyors. They had 5 partners so if one left and one was ill at least there was someone from a pool of three who could run the Green Street Retrofit.

Someone who knew someone in Didley Building Society said may be they could lend the money. How much was at that stage unknown. Kate and Quentin took advice and decided work at each house might cost £20,000. Adding for fees plus VAT[5] worked out at £30,000 so 50 houses meant a budget of £1.5 million. It did seem that the number of houses involved brought down costs per property to a useful degree.

In fact, they kept with Solemn Surveyors as contract administrators. Using documentation, they had produced they went out to tenders from a list of medium size builders.

But others had to be involved before this stage. Insurers were considered to be a crucial part, and two kinds of them. It was not so much fire or flooding and in any case each house owner had their own insurance for these and other perils. It was to cover the unexpected. Another aspect of insurance was to monitor designs and to provide warranties.

Though their 50 houses were all built at the same time around 1900 some addresses were known to have had owners that had not maintained them well and some dubious alterations had been carried out in some cases. Also, subsidence had affected the rear kitchen wings of two houses near the canal.

Between themselves, having taken advice, the committee sought out warranty insurers based in the city of London. They had been assessing quality for a number of housing

---

[5] See comments later on VAT.

association projects. They had their own technical consultants. The hope was the warranty assurers would provide at least 15 years of cover once the work was done, much as the builder of a new house may provide 10 years assurance through their national body.

CHAPTER TWO
– possibilities for the retrofit club set-up

- Scenes: the retrofit club; funding mass retrofit; insurers; retrofit builders; life plan approach; surveying and assessments; houses for the 22nd century; government role; approvals; search and remediate; condensation; full assurance; insurance risks; nerves of steel; other professionals; building wise – need for remediation; other unexpected deficiencies possible.

The retrofit club

To get beyond mere talk of hopes and aspirations a retrofit club will need to agree its constitution, its rules. One approach may suit a club in the city, another a club in the suburbs. The club will need to choose office holders, a chairperson, treasurer, secretary, probably a technical committee. To be a legal entity it will need to register and submit annual accounts. Initially these may be relatively minimal and related to the number of members.

The club will have expenses so it will cost to join it. Best this is for each house since costs of retrofit should relate to the building not an individual. Then when the house changes to new ownership all the arrangements it has are portered on. Possibly there will be associate membership for residents who are not owners.

Maybe there will be an extra annual fee of £100 for a house to be a full member once commercial activity is in mind. Other fees are likely to be payable as more complex commitments are taken on by the club. There might be an annual subscription too, but this could be relatively nominal.

To assist old people without much income a bursary scheme could be considered, and possibly a supporting charity. It may be possible to put charges against the property at the Land Registry.

The club will have an annual general meeting. And why not a garden party in summer? And each winter why not have another party? One for kids, one for others.

Funding mass retrofit

Costs of running the club are small beer. To do the job, to do 'the works', money has to be borrowed from the future. Don't forget, all this is for easing the danger to planet Earth due to the climate emergency. Ideally none of this should be necessary, but then…

To have a retrofit mortgage perhaps the best way will be to engage with and help found a retrofit building society. Like former building societies, some of which were taken over by banks in the 1990s, a retrofit building society will pride themselves on being local and of what they can provide. But some banks may be equally forthcoming and user-friendly.

One of these lends the retrofit building costs and possibly those of the necessary consultants and others, to the club, apportioned between houses if appropriate. Over subsequent 10, 15 or 20 years, they recoup, borrowers paying amounts similar to what their heating bills would have been without retrofit.

Once a retrofit building society or bank equivalent is in place and perhaps before the retrofit builder starts, most out-of-pocket or initial fees might be packaged up up into the mortgage, so some subscriptions and fees may come back. But for things to start, at least some outlay is required of a member to register they recognise what is happening, that they are part of something big.

The old and infirm may be eligible to funding via the Energy Companies Obligation – ask the energy supplier. And if a government is serious about climate change new 'green deals' may be on offer.

Insurers

Other than the owner occupying the house and their family, no one is more important in mass retrofit than insurers. They are present to cover a variety of things not working out to plan and in more subtle ways.

They are sometimes clever insurers who will have departments of building professionals to monitor retrofit methods proposed. They are providers of *warranties*.

But other insurers are aware of human frailties and cover these. For example, a house due for retrofit next week has to be displaced till next year as a resident has serious illness. Or someone has died and those remaining will need to accustom themselves to living without them. Or if the agreement allows, because a close relative (parent or child) is dying.

But most of all insurers will look across the future and offer provision to tackle the worst outcomes including some unknown at present. I look further at insurance later.

Retrofit Builders

They are not just builders, they are retrofit builders. They also empathize with what happens in people's homes. Basically, everything and anything.

You may have heard of them: Tesco, Sainsbury's, Waitrose, the Co-op, Asda etc. This book encourages these supermarkets, their like and others to set up retrofit companies. Then people will have a choice as to which retrofit supermarket is commissioned, just as they have a choice in where they buy their food. In effect a vote of
those in your street will decide.

Life plan approach

We advocate a *building life plan* approach. This will include assigning an anticipated life span to all elements dealt with and installed and the house within which these exist. It will cover costs of renewing retrofit works that fail to last, and financial planning into the future.

But it will need to go further. On the basis that cracks and other defects cannot just be covered over, any such deficiencies will need to be put right first. This will be in a concerted way. So, there will be some assurance of quality regarding the existing but remediated construction.

To do this special warranty insurance must be in place and quality assurance inspectors will have to check works. This will be the return of the Clerk of Works![6]

---

[6] Clerks of Works used to be present in most building projects.

You may wonder 'doesn't local authority building control check quality?' Well to a growing extent their checks are compliance with health and safety requirements, that tends to mean health and safety immediately or say within two or three years. Also, in theory they check compliance with thermal performance. And that was a thermal performance which is far from net-zero and was far behind what many aware people called for.

Some developers however did not want to be behoven to the local authority, so the government made provision for *independent inspectors*. Other than notifying the local authority of their project's details independent inspectors form an opinion as to whether Building Regulations have been correctly followed. They possibly can get poor and unsatisfactory work re-done. However, they might not get another contract with that particular developer or builder again if they get too awkward! You may wonder if they really are independent!

This book talks about warranties or Enhanced Retrofit Passports elsewhere.

From the inquiry into the Grenfell Fire we know at least some local authority building control offices are hopelessly understaffed and overworked. In early 2023 Secretary of State Michael Gove admitted that "faulty and ambiguous" government guidance was partly responsible for the Grenfell Tower tragedy. He said lax regulation allowed cladding firms to "put people in danger in order to make a profit"[7].

Employed by the client they checked workmanship was correct. If something odd happened, they called in the designer. The government got rid of Clerks of Works in public sector housing schemes in the 1980s. It was almost as if they said 'look how clever we are, we have saved money!'

Surveying and assessments

First a measure-up! This will use methods old and new. Old is a steel tape measure and optical instruments set on tripods. Pretty new is *photogrammetry* and automatic theodolites. For retrofit this book advocates an 'eye-in-the-sky' drone camera (which visits externally and internally including lofts) together with other more traditional surveying methods all processed by computer programmes. This will provide a repository of all significant dimensions. If required, this three-dimensional information can be printed in two-dimensions and will look like conventional construction drawings.

Thanks to the photogrammetry there will be knowledge available at the push of a button of not just wall positions but, for instance, out-of-true elements. This will include leaning or bulging walls. 'Artificial intelligence' (AI) will in effect learn surveying and what a house is. It will be able to present a list of, say, missing walls in the whole street (for instance places where a 'knock through' was done). Or a list of bulging or out-of-plumb walls, including enclosed walls such as party walls between different ownerships. Or a sagging ceiling. Or a crack in a rear wing which subsided into soft earth beside the canal perhaps.

Bulging could indicate the two halves of a 230mm[8] thick brick wall are becoming detached, are bursting apart. Walls can be

---

[7] https://www.dailymail.co.uk/news/article-11688961/Michael-Gove-admits-faulty-ambiguous-guidance-allowed-Grenfell-tragedy-happen.html
[8] Metrication: house building became metric around 1970, the National Building Agency guiding with a unified system. More on this in an Appendix.

out-of-plumb because they rely on lateral restraint being provided by floors above ground level and this can be insufficient.⁹

It is to be hoped the AI 3D camera will also pick out significant cracking in brick walls though often plaster and wallpaper may cover this.

This technology may not be available for early retrofit projects. These trail-blazers may have to rely on surveyors with tape measures, plumb lines, one metre long spirit levels and clipboards!

With this knowledge judgements can be made by specialists regarding remedial works needed to walls and relating to other defects. There will be a strong wish not to need to demolish and replace walls. Obviously, a home is for a time no longer a home if walls must be replaced. The hope will be to use special wall ties, some of types yet to be developed, some already in use. There will be other approaches to compensate for inadequacies in walls or other elements.

---

⁹ It should not be assumed that only modern building quality can be below par. If the bricklayer had insufficient lime on a Saturday in the 1880s and there was a football match he wanted to get to, he might have made up his mortar with just too little lime. The author once found this in upper parts of party walls in Peckham. Look, the builder said, you can just lift off these bricks with your hand! (The plan was to add a storey to get extra bedrooms. The foundations had already been strengthened by underpinning.) So, the engineer said, put stainless steel bed-joint reinforcement in every second of three bed-joints as you raise the new storey. This provided the effect of a continuous load-spreader and tie along the wall. Hopefully this wall will now outlast the 2080s!

Houses for the 22nd century

Your author makes no bones about this. As a structural engineer called by surveyors to inspect houses, and having done this in some detail for around 2,000 houses, his opinion is that insidious defects will worsen and become significant within 60 years, and more than likely sooner, if not addressed and suitably remediated at retrofit stage.

It should be born in mind that many houses in inner London were built before 1880 so by 2080 will be over 200 years old. It seems unlikely that there will be moves to re-build large zones of inner London. So as year 2099 comes nearer and approaching the 22nd Century, this book advocates that a new, cultivated, approach will be called for, warranties that the house was validated when retrofitted being provided. Such warranties or *Passports* may well add value to the house.

Government role

It may be noticed that I do not put high responsibility onto central government. However, I do expect it to recognise that the national housing stock is such a vast and valuable asset that it should be conserved properly. So it should look to owners to have an informed policy for keeping houses in good condition long into the future. A corresponding policy should apply to houses owned by local authorities, social landlords and private landlords as well as owner-occupiers, to whom this book is primarily directed.

The author used to work for a national building authority. Though he recognises governments can make mistakes, he advocates the setting up of a new *national building agency* which forms an informed technical overview of the nation's housing stock related to upgrade as advocated in this book. A section in the Appendix says more on this.

He also advocates the setting up of *retrofit universities* to train and monitor methods amongst practitioners. To a degree these will be familiar with building practices that have affected regional house building for approaching 200 years. As stated elsewhere here, retrofit will be a new industry and for a significant time it will be larger than the house-building industry.

You see, this book does not bow down before market forces as some do! As the writer understands it from an old London School of Economics friend, even Adam Smith thought there were facilities such as ports which were to be built by the King. He fears that it is largely due to following market forces blindly that human life on planet Earth has been put in jeopardy by the climate emergency.

Approvals

Provided the house to be retrofitted is not listed as of architectural or historic importance and provided it is not in a conservation area, planning consent may not be required. It is however always advisable to check with the local authority planners.

To a growing extent it will be useful for government to provide guidance between the three competing stand points: conservation, disability use and thermal insulation.

Approval of the local authority building control office may be required if they classify the changes as 'material'. It will be advisable to check with them first. Remediating something for which their approval should have been obtained by others earlier, for instance altering a roof structure, probably needs them to consent to drawings and calculations which will need to be prepared and submitted and then implemented.

It is not clear whether the Health & Safety Executive would require risk assessments. It would be surprising if they weren't considered necessary since changes, sometimes complex ones, are proposed in houses within which people live and close to which people come and go.

Search and remediate

So, search and remediate is a key retro-fit policy. That defects have to be searched out, assessed and where necessary remedial measures taken. Only then should retrofit commence.

For an example consider a cracked brick wall. Cracking can be either stable or unstable. An example of unstable cracking is that due to 'modern' subsidence, that is in say subsidence in the last 20 years or so. Then specialist stitching repair will be required together with, in many cases, underpinning (the retrofit of foundations). Many house insurance policies cover repair of subsidence damage including for the costs of underpinning and as is possible, alternative accommodation, except for an excess amount, often £1000. A tree growing larger as it matures in an area of highly shrinkable clay and close to a building is a not-infrequent cause of domestic subsidence.

In some cases of cracking, an engineer may decide that stitching alone may suffice. The sort of stitching in mind uses twisted stainless steel wires set in every second, third or fourth brickwork bed joint.

There was a time when one could phone local authority building control and ask if an address in its area was on highly shrinkable clay. Now you may have to commission an expert, but it is worth asking an insurance company first!

Other defects needing attention will include where chimney breasts have been incorrectly removed and when rooms have been 'knocked into one' without proper structural design.

Most chimney breast removal affects a party wall, yet the adjoining owner or their surveyor in present circumstances do not know of this. An appendix provides more information.

Condensation

Rising damp is guarded against in houses of the last hundred years by damp-courses just above ground level. It is therefore relatively rare.

Condensation occurs in walls with poor thermal insulation and high internal humidity. It can occur in any storey. Much of the wall may be free of this due to convection currents of air within the room – supposing that the room is heated! But these air currents are reduced at wall corners, the ceiling or the floor or relating to furniture. Mould is more likely to grow at such places and others where there are impediments to circulating air 'brushing' past.

In November 2022 a toddler died from a respiratory condition caused by exposure to mould in his home, a coroner concluded[10]. Could this lead to landlords who do not cure condensation being held liable?

---

[10] https://metro.co.uk/2022/11/15/boy-2-died-after-being-exposed-to-mould-in-his-home-since-birth-17761846/

For the house owner, and possibly this will mean their retrofit club, it will be important to get a surveyor's or building physicist's opinion.

Basically however, such damp may be fully cured by installation of thermally insulating over-cladding or lining. Once the wall has dried out, mould can probably then be cleaned off without risk of it re-appearing and, perhaps subject to a clean with appropriate cleansing agent and a coat of sealing paint, re-decoration over will be possible.

So external wall retrofit could cure mould growth on the nation's walls! Possibly much such mould is in social housing in which case the owners can read this book and will not need to form a club to manage retrofit! It may all be different for them since tenants may be offered alternative accommodation whilst work is underway.

A private landlord may however be in a predicament. Can they retrofit and expect their tenant to pay rent? Could it be that the government has to support the landlord financially for a time, say a tax allowance? Back in the twentieth century some government ministers mocked social housing providers, advocating private landlords renting and being less of a burden on the public finances. If government has eventually to provide financial support to private landlords, is this doctrine about to come 'home to roost'?

Full Assurance

Well maybe not assurance for eternity, but assurance for a long and useful period of time! Housing associations have for some decades engaged in warranty schemes, called something like *housing association property mutual* that cover their new stock against design faults, poor workmanship and may be other things. Big international insurance houses now offer equivalent protection. And many new-build houses are covered by a scheme in effect owned by larger building companies, NHBC (National House Building Council).

So, the retrofit building society can also, like the club of owner occupiers, be assured as part of the warranty system incorporated in the retrofit set up.

Insurance risks

Insurers will play a big part in mass retrofit. Insurance may well be categorised into one of two types.

Firstly, insurance where *sufficient number of cases are insured* that particular knowledge of any single case is not required. For instance an estate agent surveyor's insurance; they provide a written survey to a purchasing individual and this is used by a mortgage lender to agree funding. So, the indemnity insurance that a particular surveyor has in place can be claimed.

In effect, anyone who depended on a survey that missed something so causing them to lose out economically, can sue their surveyor. The surveyor's insurers may in due course take the view that they will make financial recompense to the buyer or their lender on the basis that it was a dwelling house of some sort or other in the UK and so part of the sum total of UK dwelling houses. Whilst taking account of different types, these are a known collection.

If a particular surveyor has too many claims against them, they may well find it difficult or more expensive to obtain professional indemnity insurance in future. In these ways there is protection from calamity.

Secondly, insurance where the quality of building design and construction is *monitored* and must meet standards developed and recognised as good practice by insurers. In these ways risks taken by those insuring are mitigated.

With the latter type of cover, formal warranties become possible. These are in the nature of long term guarantees, possibly component by component. We particularly advocate this type of insurance along with the protective type.

With this in mind, it may be found useful if the retrofit club constitutes itself to have a more permanent existence since it may be asked to pay an annual subscription into a warranty scheme.

But, behind the scenes, insurers and assurers will need someone to 'lean' on. These are re-insurers – big setups that insure insurers.

The various costs of insurers will have to be met by the retrofit club, both 'simple' insurance, as first described and also the second type, special ongoing warranty insurance. The latter insurer is likely to have an address on Fenchurch Street in the City of London!

Nerves of steel

To get through all this the first retrofit clubs will need to have nerves of steel. Later ones will feel indebted to them!

Other professionals

Other 'professionals' and others likely to be 'on the scene' are included in the 'cast list', see end of the Appendix. These are people other than residents who will have a role in making retrofit happen successfully - and not stopping you from making a cup of tea in your home for too long!

Building wise: need for remediation

Everyone knows it is unwise to paper over cracks and this book has already said as much.

Cracks can result from initial shrinkage for example of plaster. But they could be due to bigger issues. An example is subsidence, where the ground shrinks and walls settle into the gap. If your house has this you have probably made a claim on your insurance policy since most house insurance covers for this.

As mentioned, another category of defect results from incorrect alteration. And the older the house is the more owners it has had, so the more possible this is. Especially if there was an occupant who was an avid watcher of tv programmes about home improvement!

All in all, there are quite a few contenders in the list of possible needs for remedial works to an old house before retrofit. An Appendix lists some of these defects.

Other unexpected deficiencies possible

Issues can arise with party walls that are just one brick thick (9 inches or approx. 23cm). For instance houses built at low cost when Queen Victoria was on the throne (1837–1901) can have half brick thick *withes* (the wall at the back of the fireplace). So little more than 4 inches thick (11cm). As well as less sound resisting this indicate the brickwork bonding pattern had to be changed by the builder and was therefore not so strong. Even so there are few houses where this has caused problems, at least not yet. However, this book considers it not prudent to assume there never will be problems, particularly as 19th century housing will it seems still be someone's home in the 22nd century.

Another deficiency is surprising. In recent decades people living in such houses have wanted their new electricity circuits to have recessed power points, and not unusually in the nooks beside the chimney breast. The same logic may have applied in the neighbouring house and with similar positioning. The result: in terms of brickwork, a hole between houses – only the electric pattress boxes seal these. Therefore: not so mouse proof, not so fire resistant. If next door is on fire will smoke enter your house this way? Will your house be okay after a bad fire next door? This is another example of what eye-in-the-sky surveys together with AI should seek to identify.

Now a word of caution. These horrors may become a big issue. But house insurers can respond. They know the odds, currently considered low associated with some of the peculiarities mentioned here. And no doubt as the 21st century passes they will in effect reflect concern where there is concern. It may then cost little more to insure such house types.

CASE STUDY TWO

The Second Mass Retrofit, Thunberg Street

The second MR project members had chosen a steering group with versatilities as good as the first.

A builder was got onto site within two years of the retrofit club being founded.

Here the selection of houses was more varied: four short terraces, five pairs of semis, an old vicarage and round the corner some 1920 council built terraces. So, 50 houses in all. A key factor was having sufficient number of properties adjacent since there would be cold bridges caused adjacent to and by a house without external wall insulation.

This meant a number of retrofit designs had been deployed. They got some advice 'for free' when a local college interested in becoming a university of retrofit had supplied assistance. And a regional builder founded 80 years before, now headquartered in a former C19 industrial magnate's mansion had taken an enlightened interest.

On inviting tenders for the retrofit, *Civic Plunge* got the job. They had taken on a variety of urban renovations in an intelligent way and wanted a project to, as it were, learn mass retrofit. And they did this well, despite the continuing lack of skills due to the government not being able to make up for losses caused by Brexit.

The project completed effectively, possibly being assisted by clement summer weather.

CHAPTER THREE

-including a picture of the sorts of things your experts need to do to set up contracts to carry out mass retrofit

- Scenes: surveys; existing houses; other factors; current insulation levels; external wall cladding; a new type of gutter; fixing of cladding; appearance; features of an external wall; ground storeys; other aspects of external walls; internal wall lining; roofs and their insulation; how a loose timbered roof was built; snow; loose timber roofs in use; altering a roof; central valley roofs; trussed rafter roofs; solar panels; chimneys; incorrect removal of chimney breasts; loft insulation; a water tank; loft storage; roof safeguards; ground floors; retrofitting of ground floors; purists and wool.

Surveys

The first activity on-site, that is on your street, and well before the retrofit builder arrives, is to survey the houses. We set out this in Chapter Two. This survey will be by experienced professionals using established and newly devised methods. Eventually, after a prototype period, these will be associated with a university of retrofit.

Surveys will be both the taking of dimensions and assessment of the construction type and condition. Assessment will include fully understanding the types of houses present. Many streets and especially streets with terrace houses will have a similarity of house types. Pre 1920 houses tend to need a different approach to post 1920 houses.

Existing houses

Few people may realise that Britain is unique in having so many houses from the nineteenth century still in use and loved, albeit adapted and fitted with modern features and functions.[11]

In the nineteenth centuries most builders were small family type set ups. They would probably build two or three houses a year, broadly with the same dimensions and features as the one adjoining. This may have been finished two years before and now have a family in. Many feature items such as keystone type bricks with designs or features would have been purchased from the same builder's merchants.

There were some big builders, however. Thomas Cubitt, with his yard just off the Greys Inn Road in London, for instance. Your author once worked for the 20th century Cubitts, building Thamesmead! (Cubitt Town on the Isle of Dogs is named after someone else!)

By 1920 things were changing with big construction companies emerging. Many bricklayers did not come back from the World War. Another thing was that in houses the cavity wall started to replace the solid wall throughout. And mortar was now made using Portland cement, so higher stress levels could be sustained. Moreover, bricklaying was possible during the winter so long as the temperature was somewhat above zero. (Lime mortar needs mild weather.)

---

[11] See Appendix 7 regarding wall construction.

Other factors

As already stated, it will be necessary to repair serious defects first, as soon they will be covered over for decades. See the Appendix for more in this.

Current insulation levels

A cavity wall without insulation can have a U-Value, the measure of insulation, of somewhere between 1.6 W/m2k and 0.6 W/m2k, depending on when it was built. By installing cavity wall insulation in your home you can help to improve your U-Value and in turn, improve the insulation of your home. A realistic hope will be significantly less than 0.3 W/m2k.

External wall cladding

The ideal is that thermal cladding is added to the exterior. It may be 15 to 25cm thick and whilst largely a non-flammable plastic type product made in a factory, is likely to include metal stiffeners. These panels may be made to a stringent specification and to exact dimensions measured up from the photogrammetry already carried out by drone camera and AI processing. In some cases their inner surface may replicate in reverse the shape of the wall surface.

Good practice would suggest that cladding systems should be licenced by an impartial body following stringent testing. This is how medium rise panel building systems were permitted by government in the 1960s and 1970s. However, since such bodies were privatised, or closed[12] and since university

---

[12] National Building Agency (NBA), closed in 1981; the long

departments may rely on grants from companies, it seems the impartiality that could once be claimed is no more.

A key aspect will be performance in case of fire and any company putting their cladding on the market must be ready to demonstrate fire safety. Simulated fires, possibly at Cardington, a large former airship hanger in Bedfordshire could be the way! Yes, this will be expensive but with a target market of approaching 20 million houses the expense of getting a licence will not be unreasonable. Though unimpressed by the blind following of market forces the writer does recognise the versatility that the private sector can offer when it comes to large scale enterprise.

Of course, licencing should have been done with cladding for residential tower blocks. But some think the government was then in the pockets of certain newspapers and big builders who persuaded too many people that proper regulation was 'red tape'. The official report on the Grenfell Tower fire in which 72 died, due in 2024, may provide useful comments on this.

Different systems of over cladding may become available but a favourite of the yet-to-be-created retrofit industry may well be storey height panels, or panels slightly over a storey height, and pretty much full house width. They will likely cover over the building from ground storey window head level upwards. They will incorporate self-draining overlaps either at floor levels or possibly 1st floor window-sill levels. Such an overlap position may hark back to the platband on a Georgian house. Vertical abutments may best be made using cover-strips.

---

established Building Research Establishment was privatized ten years later.

As mentioned elsewhere junctions to other enclosing elements will be something that it is essential to achieve in a fully effective way. Examples are other walls and of course the roof.

Most houses have rain-water gutters at the front and at the rear. Exceptions include the great numbers of inner London houses built in the middle of the nineteenth century which, harking back to the Georgian style, have parapets at the front, rainwater being directed towards the rear by the central valley roof. We look at this issue elsewhere.

A big question will be how to clad up to the underside of a gutter. The big problem with gutters is they can crack or joints can open and then rain-water gets through. However, water cannot be allowed to get into or behind over-cladding.

Overhanging house gutters were developed after the middle ages, in some more remote areas perhaps more than 100 years later. Till then there were none and rain-water spilt off and unless you wore a hat, down your neck! An exception was of course buildings other than houses, for example the castle or the cathedral. Here lead sheeting lined gutters and flat areas behind parapets, with a rainwater down-pipe, sometimes internal, to drain these.

Large bracket arrangements for gutters started to be fitted to existing buildings. And from Queen Anne's time the wealthy incorporated gutters in the design. Georgians thought these rather vulgar and so used parapets to conceal leaded gutters and rain-water downpipes internally or on the rear elevation. As just mentioned, even the smallest London terrace houses built in the middle of the nineteenth century sometimes kept up Georgian appearances, the roof draining to the rear.

A new type of gutter

It may be possible to incorporate a gutter extension into over-cladding if a high level of assurance regarding water-tightness can be assured. This way the need to alter roof structures may be avoided. (Elsewhere we look at how to ventilate rooves and prevent decay of roof timbers.)

However, it is suggested here that gutters be located anew outside the plane of the new cladding possibly with brackets fixed through it into the now hidden wall behind. Such purpose designed brackets may tend to look chunky with uppermost fixings being a distance below the top of the wall (since brickwork tends not to have full strength in its top courses) and brackets would need lower fixings 400mm or 500mm below these.

Steel bolts will tend to form cold bridges. Hopefully carbon can be put to a positive use and suitable non-metal fixing bolts brought into widespread use well before 2030.

But there may be a bigger issue. This relates to the geometry of the, this being dictated by the roof slope. In some cases there may just not be enough space for sufficient thermal quilting! It may be possible to avoid a cold bridge all along by altering the roof, or at least its slope in the approach to the eaves. This could be expensive and the programme could take longer.

Another possibility to be considered could be relatively thin high-performance internal wall insulation between picture rail level and ceiling on and near external walls. The main room size so far as furniture fit may not then be reduced. Definitely run this by your Building Physicist![13]

---

[13] You may have difficulty employing a Building Physicist. As yet there are few around, so determination is needed. But they have

At this stage, one at which a wholesale change to one of the houses main features – the eaves - one might just hope the central government will step in and offer a helping financial hand for instance they pay the local authority council tax that year!

But here is scope for invention in the long term! A system for extending rafters made without metal may be devised. One can see overlapping pieces each fitted over and to a rafter, these extending outwards with the fascia fitted to it and gutter to fascia. The system might succeed by including thermal insulation and satisfying that newcomer – the building physicist, so far as dew point positioning relative to timber is concerned.

Fixing of external cladding

Possibly external cladding will in effect be glued on. But the worry will be that there will be interstices left. It is likely that a vacuum type process will be necessary to fully 'pull on' the cladding. Sonic testing will be called for to prove intimate contact.

Fifteen years tends to be the key period during which the expected and the unanticipated can happen to a building. Will, by then, minute bits of debris or mini life forms start to expand, to ratchet-out the cladding, to prize it off? Thus, systems of cladding that use a compressed layer of mineral wool or the like on the original wall surface could be more

---

a programme called WUFI originating in Germany. Input your eaves geometry into this (as it were) and get your answer: a plot of the dew-point line. This should be nowhere near timber lest it starts to rot. Mums: get your kid to be a Building Physicist!

viable. Such ways may depend on bolts being developed that do not form thermal bridges, so not metal, perhaps carbon instead. An instance of carbon behaving to save the planet this time!

Well let's go on in hope. The history of vernacular building is things that worked are still there for us to see. We enjoy trips to Cotswold villages or Bath. But ways of building done in the past that failed tend to have gone. An unsuccessful system is the last thing the United Kingdom needs.

Adding over-cladding will involve a lot of lifting. But it will not be heavy-lifting. A tower crane may not be needed! Some ingenuity may provide ways of sliding and hoisting components in almost inside scaffolding. Likewise where a tree is close. It may be possible to avoid spoiling the garden too much if sleeper type matting is put out just where hydraulic lifting trucks will stand.

Appearance

What will an over clad-house look like? Your author almost regrets raising the topic of over-cladding. He fears some results may be offensive to the eye, or in particular, his eye.

Will we no more enjoy the earth colours of brick elevations? Will lovely English houses be encapsulated? Will gaudily coloured houses be the result? Will panels incorporate incised designs which could derive from anything between Roman or Renaissance architecture (legitimate) to Disney be installed? Possibly the 'half-timbered Tudor' style popular in the 1930s into the 1960s will again be in favour. Or ship-lap boarding… Preferably brickwork that looks like the brickwork being covered, but only to the untutored eye. The potential horror is too much for your author and he will now take a rest.

On resuming... to say there should be an annual competition for good retrofit design!

Features of an external wall

A wall will of course have windows. Windows have reveals each side, a lintel or arch over, and a sill at the bottom. The reveals will need lining. Here may be a good instance for the use of thin but highly insulating panels, possibly a product yet to be developed. This may do for the window head also. The sill is potentially tricky. In purist terms it should be replaced with one projecting beyond the cladding. We will leave others to discover best ways to achieve this. This is a reason why universities of retrofit are to be set up, ideally with a national body co-ordinating methodologies - an NBA[14]. And also a reason why early retrofits may be more difficult than later retrofit.

Ground storeys

Over-cladding in ground storeys will often not be feasible. For instance, such parts of houses need high durability. They can sustain unfortunate damage if someone backs their car onto them! Or if a local young gentleman decides to try out his knife. And adjacent features and pathways can be impeded if even 10cm is added to a wall surface, for example, adjoining flights of steps down to a lower level.

So, for ground storeys, the mind moves towards internal wall insulation. This is dealt with following.

---

[14] See the Appendix regarding a National Building Agency (NBA)

Other aspects of external walls

External walls have important things attached. Take the downpipes which transmit rain-water down from gutters. These used to be cast iron, which led to a European professor of building to josh how Brits seem to hold up their roofs on cast iron pipes! Nowadays though downpipes are plastic or thin metal.

The real issue for over-cladding is how to allow for these downpipes. An initial mode of thought might be that they would have to be removed and re-mounted proud of the new cladding. But further reflection suggests cladding might be tailored to fit up to downpipes and other fittings, subject to some form of access being available so checks they are not leaking are possible.

Alternatively 'swan-necks' could be used just above ground level to collect from re-aligned downpipes and join into the existing drains.

This is an example of where after an initial period of retrofit, this new industry establishes suitable practice in conjunction with retrofit universities. After all, the insulation effect of a pipe well set into cladding might not be too poor. The tenor of this book is that we are not in an ideal situation, planet Earth being in an emergency condition. So, perfection may not be achievable but what is achieved should go a long way towards 'good enough'.

But this is not the end of rainwater disposal issues. Downpipes lead into drains in the ground. There is much to be said for avoiding the need to re-route ground drains. You may end up needing new drain runs everywhere in the street. Expensive! And messy!

Some will consider alternatives to downpipes, one being the metal chain. But budding young local athletes may mistake these for something in a gym and climb up them!

Foul drainage downpipe issues may be similar enough to rainwater disposal for similar methods to prevail. Foul drain pipes are often to be seen on the backs of Victorian houses.

Internal wall lining

Though a new discipline, Building Physicists have determined that we should not install and rely on thermal insulation to the inside of an external wall near built-in timber floor joists. In cold weather the dew point will be moved towards the inside wall surface and will 'create' damp around the timber joist bearing, these usually being 10cm long. With condensation, this bearing length is likely to decay. This is something which must absolutely be guarded against. A principle of retrofit must be Do No Harm.

Ever since The Great Fire of London in 1666 the authorities (then the King) would not permit timber in party walls, at least in London houses. Clearly thought had been given to the matter and fireproof party walls was the answer arrived at – just at the time astronomers were seeing outside of the solar system. And soon after one of these 'greats' - Sir Christopher Wren - started to build St Paul's Cathedral.

So then and ever since one-brick-thick (nine inches, 230mm) party walls with no inbuilt timber has been thought a good way to protect the adjoining house along from fire spread. We may therefore think of the Great Fire of London in 1666 as the first spur to unifying building regulation. Likewise we may come to think of Mass Retrofit as a great spur to unifying building technology for the zero-carbon ages we must plan for and hope to survive in.

So, at least in London, instead of being built into party walls, floor joist timbers are built into front or rear walls and supported in the middle of the house by a spine wall. This is usually timber originally covered in lath and plaster, but by nowadays often covered in plasterboard.

In 1990 the author lined the inside of his flank wall (at the end of a terrace) at his east London Victorian house using 75mm thick 'thermal plasterboard' (12.5mm plasterboard with 62mm expanded polystyrene backing). This worked well and saved energy effectively for three reasons:
- It was a large house, so the width of room lost was not significant, and
- The walls were 34cm thick (13 ½ inch), that is brick-and-a-half so had a 50% higher insulation value from the start;
- No timber joist was built into it, so there was no timber at the re-positioned dew-point contour, and
- The house was being fully refurbished before moving in, so there were not previous layouts of furniture to think about.

Relatively few situations are like this however. The mass of London houses are built in solid 9 inch, 230mm nominal thickness brickwork (a brick header is seen in both faces, with stretchers, that is long ways between them). And for 200 years most are yellow *London Stocks,* a brick which many think a very attractive brick.[15] As far as the general view of London is

---

[15] London Stock brick clay was dug out of 'brick fields', mixed with ash from domestic fires (and possibly other things) and fired in 'clamps'. Originally the family lived adjoining the clamp to keep warm! This clay and additions thus contained small pieces of flammable material which helped with the firing. Pores remain in

concerned these are on a par with Portland stone and Crown Estates buildings all painted the same shade of *County Cream*.

A major problem with internal lining is that it tends to reduce the size of the room. Then maybe your furniture won't fit back in!

So, yes, you may wonder about thin insulation, that is no more than 10 – 15 mm thick. You may have heard of aerogel material. With 20 million houses to insulate, a competitive market and a range of products may bring the price of aerogel in thin sheet form down significantly. Production of such material may entail emissions of carbon. But if it then keeps a house warm for a hundred years it could be a price worth paying.

So, in conclusion, this book advocates external cladding that extends from inside the rain-water gutters down to ground floor or lowest storey window head level thereby protecting timber floor joist ends from decay due to condensation.

In some cases thin insulation in the top parts of a wall, internally just above picture rail level may be successful. But take your Building Physicists opinion first.[16]

---

the finished brick, probably enhancing its thermal insulation to a degree and in other ways. Charles Dickens' novel 'Bleak House' includes a brick makers family keeping warm by living close to the hot clamp inside which London stock bricks were made.

[16] And take note the Building Physicist is a new speciality. There are not too many about, yet.

Roofs

Remembering that Mass Retrofit relates to the mass of housing which already fills English and Welsh cities and towns, it may be said that most roof structures are one of three kinds:
- *loose timbered pitched roofs*: that is a ridge and a slope down to a gutter on both front and rear elevations;
- *loose timbered central valley roofs*, the central valley is hidden by a front parapet wall, Georgian style, and draining to a rain-water downpipe on the rear elevation, this is sometimes called a London Roof, and
- *trussed rafter roofs*. These are usually triangular timber frames with infill truss members, probably provided at 600mm centres. Metal nail plates join timbers together to form triangulated trusses. They are brought to site from a factory on a lorry and craned into position. These almost totally replaced loose timbered roofs from the 1970s.

How a loose timbered roof was built

Originally, economies of roof construction, almost like Darwin's natural selection or folk music, led to the majority of house roof structures having timber members cut to length on site and immediately installed by expert carpenters. In a terrace house they comprised:
- *Rafters* often 5" x 2" at 14" centres extended from the front or rear walls up to the *ridge*. Rafters acted as a double span beam, supported half-way up by a
- *Purlin* which, in line with the king's regulations following the great fire in London of 1665 was generally not built into the party walls. Instead, they may have been supported on brick brackets or corbels. These sturdy timbers were supported by two or three raking (inclined)

- *Struts.* These were housed around the purlin halfway up the roof at upper ends and, met the other side strut on top of the *spine wall* at lower ends. In this way much of the roof weight and occasionally as much again of snow load were placed on to the spine wall.[17]
- *Top floor ceiling joists* usually provide tying across the building such that the outside walls are not subject to any destabilizing lateral thrusts.

Due to a rafter being what the structural engineer calls a continuous beam, that is it has more than two supports, only perhaps 15% of the roof self-weight and snow loading is transmitted into the external walls. This is as well since these have window openings, and bay window openings often of generous width at the front. And in the back wall often room-width openings to the rear-wing kitchen at the rear. These having less roof load was useful. For instance it meant that timber lintels used on the inside half of a brick wall flexed less and so were more effective. The outer half of the wall would be a stone lintel or, at least in London, a voussoir arch in bricks on end, fanning to the shallow curve of the window head.

Millions of such roofs were built. Their beauty lies in their economy, both of timber and labour. Interestingly I have been unable to find such a roof in any edition of '*Mitchells Building Construction*' even though at one time 90% of house rooves in the country were built this way!

---

[17] This contains load bearing timber vertical members (studs) and sometimes, over a ground storey through-room opening, diagonals creating a truss action.

Snow

It is important to realise that a house roof is rarely fully loaded. That takes a full snow load. In the past 70 years or so southern England has only experienced a few full snow loads, though often a sprinkling of the white stuff. To the structural engineer a full load is a depth of 300mm (12 inches) of snow. Across the house his results in an added load of several tons. Under such a snow load the rafters would give or flex but this would hardly be noticed. In some cases though this could tend to prize fixing nails out a little.

Loose timber roofs in use

Most house roofs in Britain were originally covered in Welsh slate. But eventually even that wears. Being pounded with hail stones more than once most years some exfoliation, some ice action and general physical wear occurred, edges near securing nails cracked, and so on.

So, many Welsh slated roofs lasted over 100 years. From the 1970s they needed replacing. Wales no longer had a slate industry able to cope with demand. Spanish slate was thought by some to be second-rate! But various types of fired clay tile were on the market, available, and much cheaper than slate – and being pushed by their manufacturers! The fact is however that a tile roof covering is heavier than the slate that had covered the roof. This clearly deterred no one and most UK houses pre 1920 now have former slate roofing coverings replaced by clay tiling, even though this is heavier than slate. Occasionally, asbestos cement slates were used[18].

---

[18] Asbestos fibres in the lungs kill. Asbestos should be removed by specialist companies using special protective techniques. In the slates on the roof it may not be an immediate hazard but will be if

But without a deep snow covering these heavier tiled roofs still stand. In the south there may have been a few days on several occasions, since tiling replaced slate, when deep snow overstressed the roof timbers. This would be for few days only, because originally thermal insulation in roofs was so poor or non-existent that snow on the roof often melted quickly and the load was gone. One of the possible risks of improving thermal insulation over top storeys is that a freak snow load will stay longer, may be get crustier with ice, and then be subject to another freak snow storm.

So, it seems likely that some roofs may get close to collapse in a future hard winter. Some may raise their eyes to read this. Well, part of global warming is more extreme weather events, so at times there will be sub-zero conditions. Expect the unexpected!

Altering a roof

Here we come to another issue with loose or individually timbered house roofs. Ignorant alteration. The struts at a shallow raking angle beneath the purlins are quite a nuisance to anyone intent on making an extra bedroom in the loft. So, at times, poorly informed people take these struts out. The roof does not immediately collapse, 'therefore ok', they think. They have forgotten possible snow load. As stated, this could amount to several tons one winter's day. But even then, failure may not occur. Hybrid effects including plating up the rafters and ceiling joists, sometimes with screwed-on plywood etc could have given un-designed strength and few if any freak collapses have so far occurred.

---

these are removed. Asbestos was used inside houses as thermal insulation but many houses are free of it.

Of course, adding a room in the roof that is inhabited is a material change so far as the Building Regulations are concerned. The local authority building control office (labco) would rightly call for a structural engineer's design, that is drawings and calculations. After all their role is to safeguard the public against freak dangerous events, not to impose red tape for the sake of it! Sadly they are too often thwarted by being overworked since their department is probably understaffed.

Note the word freak! If the most unusual thing happens and there is economic loss to someone doesn't the insurance company pay up? The answer is often they have. But in future some insurers may depend on a structural engineer's report more, not just one from a surveyor. When, however, they realise there are insufficient structural engineers who understand issues with the humble house they may do without – and fractionally increase their premiums!

Would anyone know if a roof has had members removed? From across the street an altered house and an unaltered house can look just the same. The author's experience is that insurers know a lot about what they insure but are reluctant to say. Perhaps they think it is commercially sensitive to display their knowledge, or could this also be a display of ignorance? (However, a group of insurers did set up a club that engineers involved with subsidence claims could join and pool findings.)

Central valley roofs

In this roof, used widely in modest terrace houses in nineteenth century London, a pair of substantially sized timbers, or jointed timbers, extend from the front elevation wall (which tends to run along the street with a parapet standing high), take some repose on the timber spine wall and bear onto the back elevation wall. Some 'fall' or slope is provided to the central gutter channel built over or between the beams so that all rain falling on the roof is directed to a rainwater downpipe on the back wall and into a drain in the ground along the backs of the houses.

Relatively short rafters span off of the central valley beam up onto corbels in brickwork of the party wall each side. Sometimes they almost seem to lean on the party or even end wall. In this way in London the King's command post 1665 not to build timber into party walls was followed.

Another name for a central valley roof is butterfly roof, due to its cross-sectional shape. There may be other nick-names given over the years by builders!

Getting into a central valley roof is awkward as the maximum height may be no more than 1.5 metres. Often a hatch is not provided through the ceiling of a room below. On the other hand it is usually possible to walk upright inside a pitched roof whose ridge may be a storey height above ceiling level.[19]

Trussed rafter roofs

Trussed rafter roofs took over in the 1970s. These houses were engineered to a degree that dwelling houses had not been before. They are triangulated wooden frames usually placed at 600mm centres and made with timbers 35mm or so thick.

---

[19] But take care not to put your foot through the top floor ceiling, instead your weight must be on the ceiling rafters! Before supplementary thicknesses of thermal quilting were used, these were visible, but now they may not be. Special precautions will be needed if entering such a loft.

Generally, they transfer roof loads onto front and rear walls. Each timber forming the trussed rafter is connected to the next using metal *nail plates*. These have bent edges which formed nails and which were machine-pressed into the truss timbers.

These supporting walls of course necessarily have openings - for windows and doors. So, such openings tend to need stronger lintel type beams over them than previously. (And portions of wall between openings sometimes needed extra strength concrete blocks in their inner leaves.)

Hybrid trussed timber girders were sometimes used where timber trussed girders carried roof loads on to metal brackets on cross walls. These made habitable rooms in lofts possible, sometimes in what are essentially bungalows. And could the thin coating of 'galvanising' zinc on such a bracket wear thin rendering the bent steel of the bracket to corrode? One day this will happen!

Trussed rafter roofs need bracing to prevent them buckling or 'snaking' sideways and to withstand wind forces on the roof as a whole. Their designers call for diagonal timbers to be fitted to the underside of the rafters, these taken at their upper end to a common point at the ridge and at their lower end close to front or back walls where party walls join. Typically, these will be 'four by ones' (nominally 100 x 25mm timbers[20]) and splice jointed half-way up. (This means the timbers overlapped the next one to the side by two or three rafters to which they were connected.)

In modern terraces with trussed rafter roofs wind bracing was originally only called for in every second house. Since

---

[20] Actually to allow plaining, 97 x 24mm. Most timber used is a small amount smaller than its 'name size'.

Westminster government brought in the right-to-buy, good practice would be bracing in every house. Indeed, central government could have chosen to only sell 'right-to-buy' houses on the basis that the then owner would have bracing installed, if it was missing. But perhaps I am too unrealistic!

Instead, here we have another example of sub-standard construction not being known – has a house that was right-to-buy had its roof-structure upgraded to provide bracing independent of the adjoining house? This is further instance of how the mass-retrofit of houses will include quality insurance in future. If your trussed rafter roof is not braced the warranty will expect it to be provided.

Sometimes unduly large quantities of bracing were fitted, therefore mitigating economies of timber that trussed rafters are supposed to provide! This was a case of over-enthusiastic carpenters, and possibly a poorly thought out bonus scheme!

Broadly speaking trussed rafter roofs cannot be altered unless a major rebuild takes place. They can also be vulnerable to unauthorized alteration. An instance could be a plumber wanting to fit a ventilator through the roof. They just cut out some timber that was in the way!

This chimes in again with concern that older houses can have hidden defects made by keen but poorly informed people. Arguably any house more than 40 years old should have a thorough expert engineer's inspection. This might entail some degree of opening-up.

Solar panels

Will solar panels be too heavy for your roof? The answer is probably yes. In cases where slate coverings have been replaced by fired clay tiles there is already a degree of overload. In that case it may be necessary to add steel beams below panel edges in the loft spanning party wall to party wall. Certainly, the best answer here is to ask your insurer if their policy covers a house with solar panels and see if they have any provisos. Eventually solar panel designers may get round this problem by including integral support beams that span between party walls. But then you may ask will a party wall award be needed? And in purist terms the answer may be yes. But really, we are trying to stop planet Earth from frying, so some 'niceties' may have to give – and perhaps laws ought to be brought in recognising this. Alternatively, the retrofit club rules can be devised accordingly.

Chimneys

Houses built before the 1960s would have brick chimneys with stacks and pots, often proudly displayed above roof level. Each chimney stack might contain two or more flues serving fireplaces. Not uncommonly however there could be up to six, eight or more. In addition these hollow columns of brickwork acted as buttresses to the party walls, so tendencies to collapse in high winds along the ridge direction were guarded against.

With gas fired central heating, chimneys became redundant but about then trussed rafter roofs took over from loose timbered ones in new housing.

So, rafter bracing just described provided this stability. But many practitioners like a well-built loose timbered roof more!

Removal of chimney breasts

This has often been done to obtain more space in an upstairs bedroom. Obviously, it is a room that was built with a fireplace, that is a hearth, fire-back, flues and a chimney piece. This applies to practically all houses built before the 1950s though some less expensive houses may have had some rooms without fireplaces. Rogue removal of chimney breasts however happened more to pre 1920 houses built in a softer lime mortar[21].

Removal in one storey usually often entailed leaving in position a chimney breast in a room above and also the stack in the loft and above the roof line.

A conclusion may be that several things are required for this often unfortunate type of alteration:
- an individual,
- a television programme,
- a sledge hammer
- a bag of plaster to make up the wall surface, and, at the end,
- a roll of wallpaper and paste.

The individual perhaps recalls a tv programme or thinks they do. They belt the sledge hammer at the chimney breast during the weekend. Mostly it is assumed that the brickwork above will stay in position, that it will 'corbel' out. And often it did! However, unless suitable new supports are inserted a looseness of bricks can develop. Eventully portions of brick could fall out. Such could fall on the pillow below ending the sweet sleep of someone, possibly for ever. Furthermore, this type of ill-advised alteration produces eccentric loadings in the wall.

---

[21] Since around 1920 most brickwork has been laid in a mortar made from sand, lime and Portland cement which results in stronger brickwork, meaning distances from windows to doors can be reduced.

Correctly, some structural support should have been provided since possibly a quarter ton of brickwork is eccentric from the centre-line of the party wall. It is therefore unbalanced.

Removing a chimney breast in the 1950s to 1970s a builder may have installed steel *gallows brackets* thinking all necessary support was so provided. This would secure the bricks in a few courses above the 'incision' level. It may have allowed the ceiling to be made good and in one level plane.

But the eccentricities in the wall and consequent re-distribution of stresses in the wall would remain. Such eccentricities in a lime mortar brick wall can be expected, in time, to 'snake' the former plumbness (verticality) of the wall. In a cement mortar wall the extra stress levels may locally not cause damage but in other parts reduce stress levels are to be expected. In extreme cases stresses may become tensile as the portion of wall tends to lean the other way. Brickwork cannot be expected to have tensile strength, so the wall is then prone to collapse.

This may sound over theoretical. But the context is many thousands of chimney breasts, at least in London, have been incorrectly removed. So, there can be expected to be several or many hundreds that are in an unstable and even dangerous state. Will anyone know? No. But if the street undergoes the type of retrofit advocated here including sophisticated computerised intelligence checks, a list of incorrectly removed chimney breasts in the street could be printed out.

The correct way to provide support when a chimney breast is removed is usually a steel beam system: a bearer just inside the line of the breast and beams spanning the room supporting each end of it. At one end these are supported on a possibly strengthened portion of the wall which included the breast. At their other ends these two beams need correct support, which usually means into a brick wall. In some cases, the steel beam system is hidden above the ceiling. Someone stood up high with a suitable metal detector may now be able to follow the line of the 'steels'. In some cases they may be in a loft and so visible from inside this.

Many, perhaps most, chimney breasts occur in a party wall there being fireplaces each side and at each level. Count the pots, dived by two and you may have the number of fireplaces!

But will the adjoining owner be happy for such support beams to enter the party wall? Or indeed for a breast to be removed on the side way from them? They may feel they have every right not to be happy, but they cannot forbid it. Correctly the matter has to be dealt with using the Party Wall Etc Act 1996 and any new structure needed provided. And at the instigating owner's expense!

One aspect of a mass retrofit agreement by owners in one street that includes party walls may properly be that they agree surveyors acting on their behalf operate the Party Wall Etc Act 1996. Alternatively, to avoid 'purist' thought, there might be a statement that endeavours will be made not to disadvantage owners. This gets close to the need for special insurance to be in place. See a section on Insurance.

So a programme of mass retrofit can be expected to encounter houses with incorrect chimney removal, at least in inner cities where in certain decades the house may have been used in a simplistic, low-cost, way.

Yet again we say it is a basic premise of this book that defective construction must be remediated before retrofit can occur. Remediated here means either repaired or an informed decision taken that it can be allowed to stay as it is.

This way we advocate that mass-retrofit includes upgrading such that English and Welsh houses may meet expectations of future inhabitants into the 22nd Century.

Walls

If plumb, that is properly vertical, and on suitable foundations, walls can last for centuries. When sophisticated computer driven survey methods are available and done both externally and internally (including in lofts), something like artificial intelligence will assess them all for plumb and possibly other defects, even though the human eye only sees them in parts. Hitherto such checks on party walls have been laborious and expensive to carry out. Essentially someone and their boy providing a stool would have measured displacements off a plumb line.

Problems can be expected with retrofit if walls have been rendered or covered in pebble-dash.

Cement render can crack due to shrinkage, leaving portions loose but large areas remain as it were totally fused on to the brickwork. Possibly the only way to overclad will be to remove loose areas and re-render. A difference in appearance will not matter since over-cladding will hide things.

Before Portland cement came into domination lime render was used. Members of the Society for the Protection of Ancient Buildings (SPAB, founded by William Morris in 1877) love lime render with its subtle hues instead of battleship grey, its propensity not to crack ruthlessly and especially since it can 'breath' – allow water vapour through. They are eager that lime render should be well maintained.

Loft insulation

Here it is, the chapter which everyone expects – loft insulation has hitherto been what people refer to, especially politicians, when they say improve insulation, save energy.

Basically, loft insulation so far has been mineral wool quilting laid in strips between the timber storey ceiling joists. Extra layers of insulation may have been added, this sometimes covers up unsatisfactory quilting for instance if builder's rubble has got onto it, a not unknown problem.

Any existing quilting will best be replaced as part of mass retrofit.

A water tank

In former times it seems water in a roof tank hardly ever froze, perhaps because sufficient heat was escaping through the top floor ceiling to keep the water temperature above freezing point. With good loft insulation this will not happened and it will become advisable to install a small heater in the water tank to prevent its contents falling near to freezing temperature.

A tangent: were there always water tanks in roofs? This may be an old wives' tale, but it is said they were called for by the supplying water board and in several million houses this meant another reservoir did not have to be built!

Loft storage

Generally top floor ceiling joists do not have the strength to support heavy storage. Putting more than the odd moderately heavy box by a wall in the loft should be avoided. Yes, maybe, to the winter duvet case however! Some 'better quality' houses may though have had much stronger top floor ceiling joists!

Roof safeguards

In the meantime two things are to be allowed for with roofs.

Firstly: Conserve through ventilation. It is this that prevents moisture content in timber forming the roof from rising so much that it becomes prone to decay. In a house newly built around year 1990 there might be a proprietary ducting just above the gutter into the loft interior, this provided to ensure an air way and through-ventilation.

Where insulation is being added to an older loft the need to prevent excess humidity in the loft is just as important but less easily guaranteed. With some older houses this pathway for ventilation may not be achievable. Cowl ventilators installed one third of the way up each roof slope through the roof covering might then be the surest way of ensuring through ventilation.

It is part of this book's retrofit philosophy that, subject to proper maintenance, once done no major builder's work should be necessary for 60 years. Possibly 60 years is thought of as two ownerships, the first owner thinking that whoever they sell to need have no anxieties for their period of tenure – effectively they can die there without this worry!

This way a retrofit done in 2030 should be satisfactory till 2090. This book advocates making this take us into the 22nd Century! And certainly, in inner cities by then there may well be many houses that have stood, been used and been adapted, for a period of 250 years.

So what is the *second* thing? *Wool*. There are those who do not favour mineral wool to insulate lofts. They strongly advocate the virtues of sheep's wool. They think of the insulation like clothing and speak of wicking actions etc. Your Building Physicist will explain more (once they have been around a bit!).

Now, many of us have had to dispose of their favourite jumper after a visitation by moths. So many of us are yet to be persuaded that moths can be diverted away from sheep's wool loft insulation!

Ground floors

From Victorian times house ground floors often consisted of timber floor boards spanning onto timber floor joists positioned 14 inch to 16 inches apart. These were smaller in section than upper floor joists since they were supported on brick "sleeper walls", 2, 3 or more in number across each room. The ground surface between these might be earth or some simple sealing system such as furnace ash rolled tight, sometimes called blinding.

Wooden ground floors are potentially problematic as far as mass retrofit is concerned. Which is unfortunate since they can leak a lot of energy. This is largely because air circulation to guard against timber decay in ground floors was provided. This was done by perforated iron plates or perforated 'airbricks' in opposite walls and gaps in sleeper walls allowing through ventilation.

After 1920 many floors of this type continued to be built though the blinding might then be two inch thick concrete, Portland cement, the binding component of concrete, having become more widely available.

During the 20th century there was a growing tendency for ground floors made of concrete to be used. These might be concrete laid on a suitable membrane on the ground and be perhaps 4 inches, 100mm, thick. Or they might be some 150mm thick and sometimes lightly reinforced with steel bars which came in the form of mesh or mats made of usually 6mm diameter steel bars. Air bricks were often still provided.

But the circulating air beneath both timber and suspended concrete floors was at outdoor temperature. Those who could afford thick carpeting used that to cosy up their rooms, possibly laid over linoleum.

And those who could afford would have a roaring coal fire in the fireplace in a number of rooms. Many though only had a fire in one room except perhaps at Christmas when they might hope to have more than one room heated.

Retro-fitting of ground floors

Subject to the occupants not being in residence, insulation below floor boards using mineral wool fitted between timber floor joists. It might be in small slabs a bit like a famous breakfast cereal and span between ledges fixed to the joist sides. However, ground floor insulation is not within the scope of this book, which is so far as possible interior work is done a room-in-a-day. This is since the floorboards would all need lifting following emptying of the room.

However such retrofit might be possible during a long family summer holiday though this would imply moving furniture out of the rooms in question to some form of storage..

If aerogel prices permit - and the building physicist approves! – sheets of this insulation could be laid like lino on the tops of concrete ground floor slabs. This would become a new service of flooring contractors, each room taking less than half a day. Doors may have to be taken off their hinges and a centimetre or so cut or planed off their bottom edges. There is nothing unusual in having to do this.

Purists

Purists may yet take issue with this book which advocates external thermal cladding above ground storey window head levels. They may say it should not be a product derived from coal or oil. Wool is best will be their chant.

Well, if they can design insulation panels that are suitably stiff and non-flammable, and moth proof, well and good. Give your house a woolly jumper!

CASE STUDY THREE

The 200th Mass retrofit, Carson Street

The HeLO (Household Liaison Officer) had called three months before and shown footage that captivated all. The residents had formed their retrofit club 15 months before that and had advisers in place within a few months of that.

In one particular way this mass retrofit was different. It is in a tightly built Victorian street in inner London, hitherto most mass retrofits have been in city suburbs surrounded by grassed areas.

To get to the rear elevations it was necessary to take down a pair of houses, putting steel frames in their place to maintain the 'bookend' support along the terrace. Then construction machines could get behind the terrace. To get along the backs garden walls had to be removed and to avoid costs of carting these away they were used as landscaping fill, contemporary design fences being put in following the MR.

A new government agency briefed to support MR and provided with significant funding had purchased these and a year later it emerged that it wasn't essential to replace them. Imaginative landscape architects had made a hanging garden from the steel framework with allusions to a Rachel Whiteread work. This artist had some decades before made a 'cast' of a house about to be demolished, only a mile away.

A little like Georgian housing most houses in Carson Street had basements so to get to these steep steps lead down, with the main entrance up steep steps. The designers decided, controversially, to use external wall insulation down to basement window head level.

On the basis that there would be many local MRs in the vicinity, a portion of the local park had been allotted for car parking so the street was cleared. Some houses in the residential street at the far end were in the retrofit club and arrangements were in place for a single vehicle lane here not two, thereby providing builders access space.

Many practitioners for this project had been on a primer course at a university of retrofit. So, everyone was clued up on the logic of retrofit. The course had included a study of how the 88th MR had gone wrong when the builder went into liquidation.

Just as the government keen for free enterprise to run the railways had more than once temporarily nationalised one of them when it went bust, there was now a facility to bail out an MR project in difficulty. A principle was that work stopped only for two weeks before the '999' MR service took over. After all these were people's homes…

CHAPTER FOUR –

The Event - about what will be happening during the key time that retrofit work is actually underway in your street.

- **Scenes**: advance preparations; initial arrival of the builder; full arrival of the builder; climate emergency; external wall retrofit methods; umbrella for your building; internal wall cladding; a week of retrofit; related matters; draughts and windows; gutters; inside the house; non-male operatives; clean up; the garden.

Advance preparations

Household liaison officers (HeLOs) - lovely people will be recruited for these roles, at least two per retrofit project. This facility should be on hand from 3 months before *The Event* till at least 3 months after it. A HeLO will liaise with everyone necessary, including if need be not too distant relatives. They will see that numerous arrangements are in place covering school arrangements, doctoring, hospital arrangements and so forth.

The builder arrives

Possibly tea and toilet facilities will be the first sign of a builder arriving! Also, some sort of site offices and a small compound where builders equipment can be stored. But do not mistake this for the full entry on to stage of *the builder*!

By now the consultants, the managing body and the builder will have done a lot of planning. It is possible that there will be a first strike into the ground for *enabling works*. What these are will depend on local circumstances. Sometimes none will be necessary. It could though be that some new facility is needed to replace something that already exists in a sub-standard way, possibly something that is a weak link such as a service being via a route that needs improving.

In some cases, enabling works could be quite ruthless. A house might even have to be demolished (having been purchased first) to obtain builder's access to rear parts of a closely built terrace. Clearly that would be a house not in use! The builder might look out for a nearby house not in use and move in themselves, making it their office and facility. Lashings of hardboard would be pinned to original floorboards and bannisters and to staircase walls also!

Of course, the retrofit surveyors specialists will have inspected some months or even a year before the builder arrives, using sophisticated equipment including drones to inspect and measure-up houses and terraces.

The builder really arrives

After this possible interim situation, the curtain will really rise. Again of course tea and toilet facilities will be first and may need to be extended. Unless an unused house is available for the duration, big site offices and a more extensive compound where builder's machines and equipment can be stored will be called for.

If you live on a street in inner London you may wonder how this can be done. The first thing to say is builders can be very ingenious. For instance, they may find a holding site up to 2 or 3 miles away which becomes their main compound, so a smaller zone is needed close in, a shuttle of trucks connecting the two. This has been done on mega size projects for example when a railway was built east to west under London — The Elizabeth Line.

It is likely the builder will need most of the roadway for their purposes. The local authority will need to have granted permission and re-arranged traffic routes for this.

Climate emergency

In 2019 MPs approved a motion to declare an environment and climate emergency. All sides of the House of Commons agreed this, but not on what to do next or even that this must be high priority!
To really facilitate mass retrofit the government will best set up enabling legislation. But many a street full of engaged house owners can decide to act without this if the housing density is not too great. One way or another, the road and parking places will have to be freed up. This may well imply the setting up of a temporary park-and-ride scheme for residents as well as local hop-on-hop-off bus services.

External wall retrofit commences

Things will have to be got out of the way. Even a beautiful tree could be lost. Though a replacement might well but be replanted it might not be out of order to hold a funeral service for it! - All to do with the better health of Planet Earth!
Access along front and possibly back elevations will be needed. Retrofit panels will have to be lifted onto upper storeys of house elevations and joined in. Cranes that can lift to anywhere will be called for, possibly special types not currently in use. This way it may even be possible to avoid working on rear elevations with concomitant damage to gardens.

Umbrellas

Almost certainly scaffolded 'umbrella roofs' will need to be provided. These will permit work on rooves and perhaps more importantly will enable building elevations to dry out so that they are suitable for insulating panels to be, as it were, glued on to their exterior or fixed on in some other way. Such scaffolding will have to allow for the lifting in of external wall cladding panels.
This way cladding and insulation will be added to walls and other places, while you are out at work! However a builder's foreman may need brief access to houses internally during overcladding.

Internal wall cladding

As described earlier this book advocates internal wall cladding in ground storeys.

In some ways this may eventually be not dissimilar to having the decorator in to hang wall-paper. That is once technology has applied itself and thin highly insulating products for instance based on aerogel are available. We referred to this in a previous chapter.

But it will frequently mean a kitchen has to be emptied with units being removed at least temporarily and such things as tiling removed from the inside surface of kitchen external walls. This may also apply to enclosures against external walls, for example cupboards or boiler enclosure. So you will loose use of this room from around 9am till 6pm.

Ideally this will be for just one day, however this may be on the basis that another day wiring and plumbing will be adjusted. A gas boiler and various plumbing pipes will be problematic. Ways and means of partly detaching these can be explored once there is more retrofit experience. Or a new heat source is added in good time. In the case of early retrofit projects some re-plumbing may be called for.

A week of retrofit

So what will happen on your street in a week whilst the retrofit builder is present near you? How will it affect you and yours?[22] Before anything most of the cars parked on the street will have to be relocated to a pound to provide space for the builder to operate and a protected path provided for pedestrian access. When you need your car you can either go on the park-n-ride bus to the pound (no cost of course) or the car can be brought to the end of the street for you (charge your firm for this?).

The more spectacular work, the External Wall Insulation (EWI) over-cladding, will be done according to circumstances, possibly a versatile truck mounted crane or using special lifting trucks, not unlike fork-lift trucks, or a scheme that allows hoisting from structural members of the umbrella roof

Connections may be something like glue having been sprayed onto the already cleaned-up and dry wall surface, vacuum pumps achieve a gap of less than half a millimetre since the backs of each panel are a reverse facsimile of the wall profile in all its complexity of shape.[23] Quite different methods could on the other hand be devised.

---

[22] For how it will affect retired people see another chapter. It is a basic premise of MR that those with great illness will not have to move out. Insurers will anticipate costs for coming back to an address a few years later.

[23] Ideally the adhesive used can periodically (say every 40 years)

Hazards

Some may say having occupied houses retrofitted will be hazardous, with people, especially children around. Well firstly risks must be assessed and mitigated. Another response might be that anyway climate emergency is unsafe and will present greater hazards. Balancing these risks will be necessary.

Related matters

Related matters will include close fitting of window head and reveal liners. Methodologies for dealing with window cills will depend on specific cases.

As already discussed, rainwater gutters are probably best replaced and supported on special brackets fixed into the original wall through the cladding.

It is important to recognise that fixings into brickwork cannot be made too near the top, otherwise the bed-joints crack and all above tends to lift off. Possibly 450mm top edge distances must be observed and this will affect the design of the bracket, perhaps making it rather conspicuous. If metal bolts can be avoided, small tendencies for cold bridging will be prevented.

Draughts and windows

Much energy inefficiency is due to draughts, cold air leakage into the house that is so extreme you feel it. Outside doors will need to be draught proofed, as well as providing a degree of

---

be nullified, the panels lifted off for inspection and then re-applied.

insulation. Modern windows will hopefully be draught free. Traditional sash windows will be prone to draughts getting through. But there are companies who fit special draught prevention strips. The author employed one such. As well as cutting draughts the place was much quieter. Planes approaching Heathrow could hardly be heard!

Once draughts are no more, ventilation becomes the issue – homo sapiens needs air to breath, fresh air. A house built this year may well have trickle ventilation through its window frames. In a *Passivhaus* air will be changed mechanically with heat recovery. Advice will best be taken prior to mass retrofit including from a building physicist.

A *Passivhaus* design probably has relatively small triple glazed windows. It might be an appropriate policy for a retrofit club to not deal with windows but leave windows such that they can be replaced at the owners' expense and to their specification, though sealing gaps between over-cladding may be difficult. In some cases it may make sense to make windows smaller – the house will conserve energy more easily then. This implies building portions of external wall in a way compatible with the original construction. A club policy on windows will be advisable.

Gutters

Gutter design for retrofit will be ripe for the entrepreneurial company to bring forward ideas.

There will need to be eaves pieces onto which a new gutter fits and joins onto existing rainwater downpipes.

A purist attitude to downpipes may be that they should be mounted on the surface of the EWI panels. But as mentioned elsewhere this may result in a need for re-aligned ground drains. This will be such expensive and messy work it will be worth   avoiding.

Inside the house

Inside the house we have explained it will be best to have thin (but expensive) internal wall insulation (IWI) in ground floor rooms.[24] As more retrofit is done a competitive market could bring the price of such IWI down. The aim of retrofitters will be that IWI on external walls (and close by on adjoining walls) will be little more bother than wallpapering.

But realistically in a kitchen for instance it will mean moving the whole kitchen cabinet and top installation away from external walls. It may have to be recognised that only partial re-installation of the kitchen may be possible by 6pm that evening. A mobile tea making point will have been provided in an adjacent room. And on bigger retrofit jobs there may temporarily be something like a 'British Restaurant' - they had these in World War 2. These may be available only a few minutes' walk away.

Some work will best be done anew. At least some electrical wiring and plumbing may be a case in point.

Non-male operatives

Arrangements maybe made so that female house owners can call for all female operatives.

---

[24] Reasons include loss of space around the house and also that over-cladding is not as durable as good old brickwork.

Clean up

After all work is complete a professional cleaning company will make your place good.

The garden

But what about the garden? Since the builder will know they must leave that in good order there will be every inducement to minimize harm to it and plants in it.

However, it may not be possible to save plants, bushes and trees that are near the building.

Climbing plants may get special treatment if the agreement with the builder allows for this. It may be possible to hang them from special frames. On the other hand, if it becomes the opinion of retrofit universities that climbing plants will harm the cladding then they may just not be advisable at all. In which case they will have to be watched out for and restrained!

Materials used

Retrofit may not use the sheer mass of building materials that new-build does. But all efforts should be taken to move towards low carbon methods. It may not work to be too purist here, so long as improvements in manufacture is under way such that in later stages of mass retrofit carbon 'spillage' is removed or all but removed.

Regarding materials the default position may best be to use those as close as possible to what were originally used. Indeed conservation methods may often be found to be very 'green'. The *Society for the Protection of Ancient Buildings* (SPAB) may have advice useful regarding any particularly old house.

CASE STUDY FOUR

The Case of number 20 Dioxide Drive

Carb and Di's house, number Twenty O2 Drive is like most in this street - an 1880s house with a pitched roof. It had ornate ridges at one time. Now it has a plane ridge. From the other side of the road the surveyor thought this looked straight enough when reporting to Carb and Di. Seven years ago they depended on this report and purchased the house with a mortgage from the Nationalwide Building Society.

Whilst in the house, the surveyor could not lift the loft hatch, hardly surprising since he did not have a ladder to get him to height.

Odds are the professional indemnity insurance company the surveyor's practice was signed up with did not expect more than a 'walk round' inspection. If he had opened the loft hatch and if he had been observant, he would have noticed the purlins supporting the rafters half way up their slope did not have strut supports. Evidently these had got in a previous owner's way. They removed the struts whose angle was approximately that of the roof slope, laid boards on the ceiling joists, put a carpet down, and let their kids use what they proudly thought was another room.

The purlins had kept their support on the party wall corbels so, back in the 1990s (was it?) it was thought okay since no particular sag in the roof was noticed as the struts were removed.

Your author has met builders who think this satisfactory. But when he mentions that a roof has to be good enough to take half a ton of snow[25] they quickly realise it is not just an engineer being pernickety!

A principal is that, following Mass Retrofit, a house is re-certified as meeting basic requirements. A sound roof is a basic requirement.
As a consequence, No 20 Dioxide Drive needed major roof work. In this case, with such major roof work the house had to be vacated so roof repair could be properly and safely done.[26]

So, the family had to move out for ten or possibly more weeks. For many this will be less than ideal. However, the resident liaison process twelve months before had alerted all residents to the MR Warranty including provision to rehouse temporarily where major defects were discovered. Then, six months before, the eye-in-the-sky surveying device had flown around the loft to capture its quantity-quality visual appraisal. An MR surveyor had seen this on his screen and subsequently visited No 20 02 Drive (with a ladder) where he had verified the lack of struts to the purlins.

There was four months warning and it was summer so snow was unlikely.

But all this meant the family would have to live elsewhere, the package permitting a ten week rental period. It paid taxi fares for the kids to go to school. The local schools were clued up about MR and knew they would have some pupils with expected home upheavals and taxi rides. The parents too, and even their employees, since MR had become part of the

---

[25] See Appendix 7.

[26] Your author is well aware that some individuals may consider they can replace raking studs without having to empty the house. Possibly more than any other industry building has a wide social range of practitioners ranging from the reckless to the delightful.

national consciousness – everyone will experience it. Just as a no- longer-new car needs its MOT a house having retrofit gets a 20 year MR warranty.

Since Carb and Di had a surveyor's report on which they had depended, lawyers acting for the Dioxide Drive Retrofit Club made a claim on the surveyor on their behalf. After review, insurers agreed to pay up, accepting that the house would have to be empty for ten or more weeks.

But Carb and Di had an idea for this money. Instead of renting a house a mile away, they would use it for a world cruise. This way they could see the world, see Carb's sister in New Zealand and Di's brother in Vancouver. And the kids Mol, Ecule and aTom would see the Pacific Ocean and join in sightseeing of its sea monsters from the ship's top deck.

The teaching profession had long been ready for pupils taking a cruise in term time. Pupils were called to at least one lesson per week from the cruise ship by means of zoom. On their return Mol, Ecule and aTom were expected to mount a display of their holiday for their school. They told of the amazing things they had learned about the world – Planet Earth. This was a subject in itself.

One of their parents had a special job. Their firm too knew everyone would have an MR experience, just as they would need time off for their mum or dad's funeral. (Of course the possibility of someone removing to a street which then goes through the whole mass retrofit experience cannot be ruled out!)

So, some zoom attendance was required each week by means of zoom. And, having a high level job, one had to fly back to London for a few days half way through the cruise, zoom not being sufficient.

So the retrofit builder had the house to itself, furniture having been put in store, insurers paying for this as part of normal cover. At this stage some other defects had been noted and could be put right more easily and safely with the house empty.

For 20 Dioxide Drive the mass retrofit Event was certainly not boring! All the bother was worth it!

CHAPTER FIVE
- Final stages, about finishing off and follow up and how to encourage others.

- Scenes: left over houses, tidy up, follow-up, snagging, odds and ends, cavities filled, energy use monitored, passport to the future, how long retrofit will last; time for a party; caution and endeavour.

Left-over houses

Whilst the 'mass retrofit' will now be complete there will be houses left over till later. Typically this will have been due to illness or family issues.

Of course there are likely to have been some houses not part of the retrofit club, maybe they were just very out of touch. Some might have dropped out unexpectedly for instance someone died and their partner couldn't face retrofit people in the house. If this happened shortly before the main retrofit was due to commence it should be covered by insurers. One day they will meet the extra costs of returning to do a single house.

Similar tasks as during the mass retrofit will now take place on the 'unfitted' houses as they become available. This may be done by a team more like a 'flying squad'! There may only be a minimal presence of the mass retrofit builder by this stage.

Almost certainly umbrella rooves will have to be left in place (or newly provided) at such houses and probably each neighbouring one. Or they are taken down and re-erected a few years later. Again, the art of scaffolding will have reached new levels of readiness and speed for this.

Tidy up

When finally, everything is done the builder will fully leave site. This will be done quietly, it will probably not be possible to know exactly when it happens!

The street will revert to what it was with cars driving along it and no doubt some car parking.

If a house was taken down to get access there will be a 'tooth missing'! If space permits this could be opportunity for the retrofit club to build a *Passivhaus*. Maybe everyone will be curious to see one of these in their street!

In other cases, the local authority might just agree to make the plot into a garden, paying necessary costs – unless Whitehall forbids!

Follow up

As is the case with any building project there will be snagging – work that isn't entirely satisfactory will have to be made good. Appointments will need to be made to do this. This may still apply to work not needing access inside the house. HeLOs will best be on hand to arrange these and have access to central bank of the club owners house keys.

The most significant parts of the Final Act may though be behind closed doors. Or better say half closed doors since all club members will expect any related matter to be concluded correctly. Warranties or Passports will be issued for each house, the Land Registry informed of these, since they will be attached to the house so go on to a future owner, not remain with the current owner.

Odds and Ends

If your house was built with cavity walls and if your professionals have, as suggested by this book, left your house with its original external surface exposed in the ground storey, be careful not to drill through the outer leaf. Polystyrene bead fill may have been blown into the cavity. This fill acts almost like a liquid and the beads will run out through the hole like a liquid! Gradually the beads will empty out and the insulation they provided will disappear from the top downwards.

So, this must be stopped forthwith! One interim way may be to take a cork out of a bottle of wine and push that into the hole, so stopping the loss of insulation. Until a more permanent solution is found.

It is beyond the scope of this book to advise what to do with the wine left in the bottle. And of course someone not embarking on mass retrofit will find knowledge about polystyrene bead fill or any other form of cavity fill equally useful.

More seriously to say that insulation was widely added to cavities from the 1960s onwards by a company blowing in either polystyrene beads or possibly other insulating material, even strands of old clothing fabric! The owner's energy use was thus reduced, though not especially so. As part of mass retrofit this relatively simple means of reducing energy consumption is to be considered. If for instance a fill offering increased insulation values becomes available it could be worth emptying out the bead fill and replacing it. If by some means bead fill has already been lost then the cavity walls in the upper parts of the top storey may have reverted to their original un-insulated condition. A check that this has not happened is well worth doing, then making good.

An equally important item can be commenced as soon as you have finished reading this book today. Record your house's energy usage each month. This might entail writing in the number of kilowatt-hours used each month in a notebook. Or possibly use an App automatically.

Then once mass retrofit is complete these figures should be seen to reduce dramatically. After all the whole idea of mass retrofit is that your and everyone else's energy use to keep walm is reduced – to zero! Electricity will though still be needed for lighting (albeit with low energy light bulbs) and for operating all those things we expect with modern life: television, computers and all the rest of it! Indeed your house system will still need electricity even to operate the place at net zero. Ventilation fans, heaters to provide hot water etc will still be needed. But much of the time your house will generate energy to put into the public network. On balance you should be net-zero!

Concluding

Final payments will need to be paid to all parties, the consultants, the surveyors, the building physicist (though this role may eventually be found best as part of a consultant's package), and of course the retrofit contractors.

Passport to the future: how long will retrofit last?

If a good warranty scheme, let's call them *Enhanced Retrofit Passports*, is in place with a good Clerk of Works inspecting as work is done, your house will last longer than it would have done untouched! It will not be unreasonable to expect it to last until the final decades of the twenty-first century, subject to routine maintenance. Then at the start of the 22nd century, 75 years from now, there will be many people happily living in houses built originally in the 19th century -provided it has a *Passport*!

Always remember

Nothing in this book is an indication that any particular house can be retrofitted. There will be some houses which cannot yet be treated in ways referred to here. In some cases, retrofit may be better done once the total of retrofit experience is greater. New materials and ways of insulating may emerge that are best for a specific situation. Then a more capable retrofit design engineer may emerge who can succeed with difficult circumstance. All consultants should carry their own professional indemnity insurance.

Time for a party!

After one or two more AGMs the Retrofit Club could lay itself down! However, under a new guise, the 'togetherness' it produced may be thought by all to be worth considering. So the organisation might then do small things, like make that corner of land into a small garden! Or it may continue to have a role managing approvals and the like.

Caution, endeavour and kindness

As in any unusual venture caution needs to be followed. Caution at the right time, endeavour at its right time, kindness always as we are all wanting to live together happily on ***Planet Earth***.

CASE STUDY FIVE

The future of planet Earth, year 2033 onwards

By year 2023 historians have shown how climatic events and the like have shaped planet Earth's history more than mankind's quirks, though these may remain of horrible significance.[27]

After a slow start, by 2033, ten years later, hundreds of mass retrofits have been carried out by the house owners of England and Wales. These have been done not only to insulate better but to make relatively ancient houses fit for the future. This has included making them mould and condensation free. Other countries have similarly attended to how they improve their shelter and mitigate adverse effects on the planet.

Much travel by 2040 is by means less harmful to Earth. Aircraft may be different and able to store energy as they descend. People will still marry someone from the other side of the planet, no doubht, so this greener flight is good!

Much manufacturing now avoids fossil fuel extraction and soon no more of this will happen. Production of hydrogen fuel from solar panels and tall wind-mills is widespread. Other ways of storing energy are being developed. For example lagoons may be built in estuaries. By night electricity from wind-mills at night pump water into these. By day these pumps generator electricity, thereby supplementing wind and other means.

Many millions of trees have been planted.

---

[27] The Earth Transformed, Peter Frankopan (2023)

So by 2033 there is hope that greenhouse gas levels will stop increasing. The benefit of the Green Street mass retrofit was taken up by many millions of others such that their need for shelter is ceasing to be harmful to Planet Earth.

In the meantime, mass retrofit of houses in England and Wales by clubs of owners will continue to a crescendo and all are done. By 2040!

APPENDICES

Appendix 1	Extracts from the government's English Housing Survey 2019-20
Appendix 2	Volume of mass retrofit
Appendix 3	The English house
Appendix 4	Unusual house types
Appendix 5	Non-standard construction
Appendix 6	Altered houses
Appendix 7	House walls, house roofs.
Appendix 8	Measurements, metrication and bricks
Appendix 9	The 'AJ Guide to Structural Surveys' by Clive Richardson, 1985
Appendix 10	Bay windows
Appendix 11	The construction professions
Appendix 12	A national building agency
Appendix 13	Decarbonising other UK buildings
Appendix 14	The grand drama of retrofit and the 'cast list'
Appendix 15	Letter to The Guardian, February 2023

APPENDIX 1

Extracts from HM Government's English Housing Survey 2019-20

Stock profile

2.4 In 2019, there were an estimated 24.4 million dwellings in England, including both occupied and vacant homes. Of these, 15.6 million (64%) were owner occupied, 4.7 million (19%) were private rented, 1.6 million (7%) were local authority and 2.5 million (10%) were housing association homes.

The age of dwellings varied by tenure. The private sector had the highest proportion of older dwellings with 23% built before 1919, compared with 6% within the social sector,

2.6 Within the social sector, most (72%) of the local authority housing stock was built between 1945 and 1980, compared with 47% of housing association homes. Just 11% of local authority stock was built after 1980, compared with 38% of housing association homes, Annex Table 2.1.

2.7 The majority of private sector dwellings were houses and bungalows (84% compared with 56% of social sector stock). There were very few detached houses in the social sector (under 1%), and more purpose built high rise flats (36%, compared to 11% in the private sector), Figure 2.3.

House condition

2.17 For a dwelling to be considered 'decent' under the Decent Homes Standard it must:
• meet the statutory minimum standard for housing (the Housing Health and Safety System (HHSRS) since April 2006),

homes which contain a Category 1 hazard under the HHSRS are considered non-decent.
- provide a reasonable degree of thermal comfort
- be in a reasonable state of repair
- have reasonably modern facilities and services

2.18 In 2019, 17% or 4.1 million homes failed to meet the Decent Homes Standard, down from 30% or 6.7 million homes in 2009.

2.19 Private rented dwellings had the highest proportion of non-decent homes (23%) while the social rented sector had the lowest (12%).

2.24 In 2019, 2% of homes had problems with condensation and mould; 1% were affected by rising damp; 1% by penetrating damp.

(The report goes on to consider types of insulation.)

Energy efficiency

2.27 The Government's Standard Assessment Procedure (SAP) is used to monitor the energy efficiency of homes. It is an index based on calculating annual space and water heating costs for a standard heating regime and is expressed on a scale of 1 (highly inefficient) to 100 (highly efficient with 100 representing zero energy costs). Findings presented in this report were calculated using Reduced Data SAP (RdSAP) version 9.93.

2.28 The energy efficiency of the English housing stock continued to improve. In 2019, the average SAP rating of English dwellings was 65 points, up from 45 points in 1996, Annex Table 2.7. This longer term upward trend was evident in all tenures. The average SAP rating of English dwellings increased from 63 in 2018 to 65 in 2019. This was evident in all tenures apart from local authority dwellings where there was no significant increase.

2.29 In 2019, social stock had an average SAP rating of 69, higher than private sector stock which had an average SAP rating of 64. The social sector was more energy efficient than the private sector, in part due to wider use of solid wall insulation, Annex Table 2.14, but also because of dwelling type. In particular, the social sector contained a higher proportion of flats compared to private sector, which have less exposed surface area (external walls and roofs) through which heat can be lost, than detached or semi-detached houses.

APPENDIX 2

Volume of mass retrofit

1. In 2021, the value of house building was £46,545 million[28]. (Some 40% of construction).
2. Cost of 2m retrofits pa at say £23,000 each is £46,000 million. (Retrofit en-masse will be significantly less expensive than this being done individually, as well as more effective.)
3. 10 years thus makes for 20,000,000 retrofits.
4. So, the volume of retrofit is similar to the value of current house building.
5. But in the peak years of retrofit this new industry will be bigger than the current house building industry. (However, many including the writer think the industry should build at least twice of the current number of dwellings[29], upwards of 400,000 per year. This was achieved in the 1950s.)
6. So, this and retrofit would be three or four times what house building now is.
7. Thus, it would be government's task to ensure a skilled workforce three or four times what it now is.

---

[28] Office of National Statistics

[29] This implies building not just the current builders' favourites, which some might call boxes for living in, but developments equal to Park Hill (Sheffield) and Dawson's Heights (East Dulwich)!

APPENDIX 3

The English House

So what is an English House? Well, the most typical and widespread were built in cities, large towns and their suburbs between around 1840 and the present day. Saying anyone particular house is typical is of course hardly possible!

The same applied in Wales.

In Scotland things were different and there was an established tradition of living in tenements, and many lived in what the English might call a bungalow, but it had an upstairs with rooms lit by large dormer windows. This book is not about Scottish houses and tenements are multi-storey flats not houses. But people of Scotland – work out how to deal with your houses!

So there are

- *Victorian houses*, millions, which developed from houses in the Georgian period. In London bye-laws classified them according to size, so there became a degree of similarity.
- *Edwardian houses*, not millions of these, but often eye-catchingly different. Victorian and Edwardian houses were virtually all built in fired clay bricks laid in mortar made with lime and sand. (It may seem surprising, but houses and architecture is often labelled according to the current monarch!)
- *Twentieth Century (C20) semi-detached houses* – like a terrace of two, but providing a way to walk round and get to the back. These 'semis' tend to be of two

storeys, many being built between around 1930 up till present times by independent builders who then sold them to their first occupants. With the advent of semis there were three highly significant innovations:
- the use of cement in the mortar along with lime and sand.
- Soon all houses were built with 'cavity walls'. These had a two inch (50mm) cavity between the outer (seen) 'leaf ', built in clay brickwork and an inner (concealed) leaf. To a growing extent and especially after 1945 the inner leaf was often concrete blockwork 100mm thick, concrete blocks costing less than bricks. Metal ties held the two leaves apart and provided some of the robustness that a solid brick wall would have, particularly because these walls were around two inches thicker.
- Some designs were 'modernist'.

- *Local authority built housing* became big after around 1920, initially inspired by the garden city movement (eg Welwyn, Letchworth). These houses would be in terraces or semi-detached. By 1939 some authorities provided space for parking a car or a way through to garages at the back.
- *Architect designed houses* Post 1950 or 1960 architect designed houses became more frequent, developed by builders or local authorities as landscaped estates, trying to include park land, sometimes hoping to keep pedestrians and cars apart. Whilst 'low rise' flats were first built before 1900 there was a recognition that flats up to perhaps four storeys in height would also be built on an estate since by then a growing percentage of the population did not live in a nuclear family. Often architects and local authority councillors were well

informed as to what was being built in other countries and this often influenced.
- *Builder designed houses* Post 1960 especially, independent builders built houses for sale. These were sometimes influenced by public sector designs, but mor often were diametrically the opposite to public sector design, that even tended towards a widespread wish to live in a cottage, though a spacious one.

In the last decades of the twentieth century central government discouraged local authorities from building housing. Consequently, the number of dwellings built in the UK tends to be less than 250,000 a year. In the 1950s it was more like 400,000 per annum when Harold Macmillan was minister of housing and there was an attitude that the state should ensure housing was available for those who fought in World War 2 ('housing fit for heros'). As a consequence, current housing can be not dissimilar to earlier housing but tends to avoid looking 'modernist'.

What is not an English House? Flats no matter in how many storeys, are not the subject of this book though they did become popular from the 1850s for instance in west inner London using 'fire-proof' construction, that is with floors not containing timber.

APPENDIX 4

Unusual house types

Dear reader, at this late stage in our thoughts together you may wonder is author only aware of one house type, possibly the mass house built in the Victorian or Edwardian eras. Taken together with the *small matter* of World War 1, this equates to pre 1920. This thought is the sign of a good reader!

In a nutshell the author hopes much of what is here said will apply to houses built since 1920.

Here is a list of some unusual house types, in no particular order!

- *System built houses*: for instance, ones built of concrete panels, especially during the brick shortage that followed World War 2.
- *Timber frame houses* (see NBA, National Building Agency)
- *Cross-wall structured houses* – their front and rear walls were panels lifted in during original construction but only following the cross walls which separated houses in the terrace. These were essentially similar to any other *party wall* since they had to have enough mass or weight to be sound proof, or nearly so. Consequently, these front and rear elevation panels could be swapped with higher specification panels, (always subject to university of retrofit approval!).
- *Architect designed detached houses*: The architect might consult a building surveyor or in particular a structural engineer for beam and foundation sizes. In fact, someone like your author in a new incarnation! If

your street is mainly detached houses you are probably reading the wrong book. The possible economies of scale etc may not be present in a street of detached houses of varying style. Nevertheless, they should be retrofitted! A retrofit experienced architect should be used.

- The *cottage orné*. From John Nash during the 1810 regency period the 'ornamental cottage' has been popular with those who can afford the expense of looking simple and wanting a thatched roof! Mostly these are 'one offs' but there is a small village of them let out by the National Trust at Blaise Hamlet near Bristol.

APPENDIX 5

Non-standard Construction

Much of what this book says relates to standard construction, that is external loadbearing masonry walls.

Various forms of non standard construction exist. Instances include:
- Following World War 2, single storey '*prefabs*' were devised as a quick way of providing houses, probably the factories formerly used for fighter aircraft for the RAF being adapted. They became more popular than the government could have expected, especially to lovers of gardening, because they were detached or semi-detached. Before World War 2 prefab residents probably lived in a very sub-standard dwelling.
- Two storey prefabs. Some of these were also 'system built' rather like single storey prefabs but possibly from thin panels containing asbestos[30] hung on steel frameworks.
- Improved prefabs, possibly called *rebuilds*. In the late twentieth century attempts were made by some local authorities to improve prefab houses which remained popular. Thoroughgoing methods involved providing new perimeter foundations and building a single thickness of brickwork externally. This was often possible as these houses were spaced out and had land

---

[30] Asbestos is bad news. It causes asbestosis, a non-curable lung disease. From around 1970 it has been banned. Commiserations if it is found in your house. But let's be realistic, a high proportion of houses have Artex coated ceilings, complete with wiggles – and asbestos fibres! And no bodies due to Artex!

around them – the garden! Ideally any panels containing asbestos were removed.
- Terraces with cross-wall construction, meaning that cross walls are loadbearing whilst front and rear elevations are panel construction and only support their own weight. Fitting new re-fabricated panels that have high insulation value may be relatively straight forward. Often such construction was built before *right-to-buy* legislation and before the local authority or social landlord knew that such elevations were expected to need replacement after, say, thirty years. Whether that was made clear to each right-to-buy purchaser is not known!

APPENDIX 6

Altered houses (see also Appendix: Non Standard Construction for rebuilds)

Most houses are altered houses. Alterations can be
- To unite rooms,
- For changing circulation routes,
- To enlarge rooms, (Removal of chimney breasts is a special case of obtaining more space. The reader should consult the body of this book for this since it was often carried out without proper regard to structural considerations.)
- To change window or door openings.
- Other sorts of alteration not primarily affecting walls.
- Extensions, but since the rest of the house may be unaltered we do not class these as alterations, except no doubt that an opening in a wall will be required to join the new to the old part.

So, after say 20 years unaltered, houses tend to consist of what the original builder left and what a variety of other builders have done to the place. Some of these may leave problems. The concept of mass-retrofit may be spoiled if the house cannot be considered adequate following it, not matter what it was like before.

So it seems mass retrofit implies a health check, at least structurally, on the nation's housing stock. By 2099 a third of this could be well over 200 years old. So the opportunity for the most valuable national asset – its housing stock – to be monitored and corrected should be welcomed by all, including politicians. Only the means of doing this could be contested.

The reader may detect that this writer offers a de-centralised way of both retrofit and quality assurance into the 22nd century.

APPENDIX 7   House walls and roofs

In Victorian times houses had solid brick walls. Such walls with just one storey over will be a nominal nine inches thick laid in lime mortar (230mm approx. or 9 inches, the length of a brick, plus allowances. Many practitioners with existing housing find it useful to think in imperial units, that is feet and inches.) Houses with three or more storeys may well have 340mm thick walls in lower storeys (13 ½ inches).

Lime mortar was lime and sand together with sand. This mostly was highly alkaline. In some industrial areas it included ash and could be highly acidic. Ash has other properties with the result that a Victorian house in the Sheffield area and probably other once industrial areas is a 'Faraday Cage'-its walls do not allow mobile phone signals to enter!

From around 1920 house walls started to be built in two 'leaves' 4 ½ inches thick (11cm nominally) with a cavity two inches (5cm) wide laid in mortar containing cement, sand and lime, not just sand and lime. The outer leaf was a facing brick in stretcher bond (all bricks show their sides only). The inner leaf would originally be a common brick, costing less. Nowadays however the inner leaf is often concrete blockwork, this being less expensive still. Across the cavity embedded at each end in the bed-joint mortar are metal wall ties. These walls, whilst not perhaps quite as robust as solid ones, have increased thermal insulation. From the 1960s expanded polystyrene beads were being blown into the cavity to improve thermal insulation. Beware drilling into such a cavity – the beads drain out like a liquid! By the 1970s sheets of thermal insulation were being incorporated into the cavity.

House roofs are expected to support 15 pounds per square foot in plan of snow. (I can't remember what this is in metric units, but do recall snow load is half of what a floor should take which is 30 pounds per square foot). In plan the roof is say some 1000 sq feet. So a full snow load is 15,000 pounds. This is approaching seven tons as there are 2240 lbs to a ton. (Note that an imperial ton and a metric tonne are, by chance, very close to being the same.)

In the case of a house in London there may only have been one full snow load each other decade since 1947. In the north of England and Wales, twice or more a decade snow will have settled to depth, possibly weighing almost up to 7 tons at an average house.

It would have usually melted within 48 hours since old loft insulation was fairly ineffective, permitting warmth from the house to get into the loft and to the roof covering. An astute observer in a house opposite might have seen this roof loose its snow quicker than its neighbour's which had better insulation.

It is not unusual to see struts removed in inner London houses. Then small people could walk about. So then purlins are being asked to support across most of the house width. With no snow, this they may do any sag not being obvious. Following a blizzard on top of settled snow, and without the support of struts, and the protection against twisting that their housing provides, they could flex badly possibly delivering a lot of snow into the bedroom!

APPENDIX 8

Measurements, metrication and bricks

In the 1970s, days when governments were not short of confidence, they got their National Building Agency (NBA) to supervise the bringing in of metric dimensioning of public sector house construction. This was to replace 'imperial' units of measurement. Soon after private house builders followed, then the rest of the construction industry.

The Systeme International was adopted, so two units of length are used, the millimetre (mm) and the metre (m), a thousand millimetres. (Many thought this scientific based system inappropriate, preferring centimetres and metres!)

There are 12 inches in one foot and three feet in one yard.

An inch is 25.4mm, one foot is 304.8mm.

In the 1970s there was *dimensional co-ordination*. Normally, not just any number of millimetres would be used, there would be preferred ones. As well as tens of millimetres there would be 25s. So, the inch (25.4mm) as it were got shorter, thanks to a system devised in France under Napoleon's instigation!

But products changed in dimension too, if only to a degree. Notable was the brick. They used to be 3/8 inch shorter in length than 9 inches, 3/8 inch shorter in width than half that, 4½ inches and 3/8 inch shorter in height than 3 inches. (The 3/8 inch was the thickness of mortar the bricklayer built the wall in.)

Brick manufacturers agreed to change their brick sizes! They would be 10mm less than 225mm long, 112mm wide and 75mm high.

Of course, bricks have their own tendencies. Clay bricks are fired in a kiln, are then brought out for a period to cool to everyday temperatures, then for a further period to settle into atmospheric conditions. They do this by taking up atmospheric moisture and in so doing get a tad larger - possibly a millimetre longer. If they are used before this process is complete the wall may crack and have to be replaced or repaired.

So metric bricks are slightly smaller than those made prior to 1970. This is most evident in the height of four courses. After a date in the early 1970s this became 300mm. Previously it was about 305mm or twelve inches. So, a proper surveyor can use this difference in dating a house!

Not to be outdone, brickwork in Sheffield and possibly some other northern parts was for a century laid with bricks such that four courses were 13 inches high. The bricks weighed the same though! The frog was just deeper, the frog being a depression on the bottom of the brick. Or was it on the top of the brick. We are now getting near to folk-lore!

Perhaps surprisingly bricks have tended to be this size since time immemorial and across Planet Earth. The reason being that a bricklayer would pick up his brick, give it a quick toss in the air to choose the best side to lay downwards, then place it.

A brick weighs around 2 kilogrammes (4.4 pounds), the weight a brickie could flip into the air ready to bed properly.

The most notable exception to this sweeping statement is the Roman brick (one does not argue with a Roman). We would tend to call their bricks tiles because they were mostly shallower (though probably weighed 2 kilogrammes still!). The result? - their brickwork looked lousy, unlike English brickwork. But that didn't matter as any wall to be seen by a person of status would be clad in stone (probably held on by metal clamps) so that it looked, they thought, like ancient Greece.

It may seem odd in the 21st century to think in imperial units but to a *practitioner* it can be very useful. If they see a nine inch thick (230mm) wall, that is a brick length is the wall width, they call it a nine inch wall, not a two hundred and thirty millimetre wall. Likewise, a chunky thirteen and a half inch brick wall (the bond pattern is based on a stretcher length and a width).

And it may be a carpenter talks about a '4 by 2' that is a piece of timber 4 inches by 2 inches. Certainly chippies (carpenters) in the USA spend half their time installing 4 by 2s!

APPENDIX 9

From the 'AJ Guide to Structural Surveys' by Clive Richardson, 1985

This booklet is a compilation of articles on structural survey written by structural engineer Clive Richardson, and published in the *Architect's Journal* in 1985. Its data sheets describe various problems and defects then known to occur in UK housing.

1. General Problems
    a. Spreading of high collared rafter roofs
    b. Rotten joist ends
    c. Stability of out-of-plumb walls
    d. Distortion of irregular plan forms
    e. Stability of untied flank walls
    f. Inadequate restraint of distorted walls
    g. Sulphate attack of mortar
    h. Partially removed chimney breasts
    i. Differential settlement between old and new construction.
2. The Industrial Revolution
    a. Snapped-header brickwork
    b. Bouncy timber floors
    c. Decay of continuous timber lintels
    d. Rotten bonding timbers
    e. Unbonded party wall and external wall junctions
    f. Unbalanced 'butterfly' roofs
    g. Falling brick arches and backing lintels
    h. Improperly formed openings in trussed partitions
    i. Lateral instability of shopfronts

3. Common Problems 1850-1939
   a. Overloaded walls
   b. End-of-terrace buildings acting as bookends
   c. Differential settlement of part basements
   d. Deterioration of filler-joisted clinker concrete slabs
   e. Corrosion of reinforcement in concrete
   f. Rot and creep of timber bressumers.
4. The post-war building boom
   a. Untied precast concrete floors and roofs
   b. Inadequate protection of walls from rainwater
   c. Lack of wall movement joints
   d. Corrosion and buckling of wall ties
   e. Eccentric lintel bearings
   f. Unsuitable bricks in parapets and below DPCs
   g. Trench fill footings in shrinkable clay soils
   h. Movement of concrete flat roofs
   i. Unbraced gable roofs and untied gable walls.

Other problems have come to light in the last 40 years. These include:

- Too-thin internal leaves of cavity walls (in the 1950s authorities experimented with public sector housing. They tried three inch (76mm), not four inch (102mm) blockwork inner leaves.) Some of these soon failed, for instance exhibiting horizontal cracking in upstairs inner leaves where rigid wall ties caused lifting due to the brickwork outer leaf expanding upwards. The bricks were used too soon after coming from the kiln and expanded a tad as they took up atmospheric moisture.
- Concrete gutters (eg 'Finlock') that contain steel bars which have corroded.
- Incorrect wall removal: the steel beam used was not strong enough or the intensification of loading on

nearby walls or foundations resulted in overstress and possibly local subsidence.
- Original bay windows being replaced by uPVC frame windows without proper attention to supporting the roof, or worse, the 1st storey bay above.

APPENDIX 10

Bay windows

Bay windows were a glory of the English terraced house and the semi. How to deal with them is the question.

It is to be regretted that in recent decades many bay windows framed in wood have met their demise. In the 1970s most still had their sash windows and corner mullions that contained heavy weights and ropes so the sashes could be easily opened and closed by moving up or moving down. By 2000 many had gone, to be replaced by uPVC framed windows sometimes with no strong support in lieu of the sash-box which, unintendedly, provided support for the roof over. Now hardly a sash box bay window is left and the prospect along a terrace of houses can be gruesome with variable detailing and some replacement windows already now calling for re-replacement! Would that there had been red-tape to ensure some degree of consistency in appearance! Even newspapers, who in what might seem some half-perverted way think any regulation or guidance is red-tape, would have articles on Saturday mornings about streets that had come through the demise of sash windows, if any had, in a pleasing way.

Sometimes, especially if there was a bay window over, builders of higher quality of houses recognized that a bay window should have some degree of stability and so they built brickwork piers or even provided stone corner mullions. This type of bay window survives! But are the mullions acting as cold bridges?

With mass retrofit perhaps the best thing will be to replace all bay windows in the street with a new prefabricated types made to measure using motor car technology and made on production lines! The hope, as ever, will be that good design prevails.

APPENDIX 11

The construction professions

A long time ago there weren't really construction professions. Either someone built a house, or a king built a castle, or the monks set to and built a monastery. The most amazing buildings had two ingredients: a lot of money, and, a genius designer. The dome on Florence Cathedral is a case in point and Brunelleschi designed it.

Nearer home a genius in what was then called natural philosophy or the like was engaged to design London's Cathedral – St Paul's. Earlier, he, Christopher Wren had looked more at the stars. Even nowadays the person who keeps a watch on St Paul's and is ready to monitor it is called the Surveyor, though they maybe a structural engineer.

Architects
But by the 19th century disciplines were getting more specific. Though the Royal Albert Hall was designed by a military engineer most other London buildings other than houses were designed by an architect, unless of course if they were a railway. Not unusually the man who designed the steam locomotives had something of a say in the construction engineering and sometimes the station too.

Civil Engineers (called that originally as opposed to Military Engineers)
Road building became a big thing especially once Irish MPs sat in the Westminster parliament. Thomas Telford designed a road all the way and a large bridge too when it came to the Menai Streets. When the Institution of Civil Engineers was founded in 1837 he became its first president.

Surveyors

Surveyors of various sorts helped the industrial revolution especially when canals were being built prior to railways. And you probably employed a surveyor when you bought your house. You *depended* on his survey. If he said the place had big problems and you bought it, perhaps this was not your finest hour! If he said the place had no problems you felt confident in buying it! Quantity Surveyors measure up building work when it is to be paid for.

Structural Engineers

Some people want to alter their house. This is not necessarily out of vanity. The children will one day want a room of their own. There is quite a lot of extending of London houses designed by architects for this reason. Further to their advice they also employed a structural engineer. Your author knows, he was one! It was not unusual for the rear wing to have an extra storey added.

Builders

Some builders just build. Some set up their own professional body: the *Chartered Institute of Building*. Others and their peers considered themselves *Master Builders*. If yours was a large ambitious company the main thing was to get bigger so joining a profession came second.

Party Wall Surveyors

Party Wall Surveyors aren't really a profession, you probably have to be one of the above first. They exist because between houses in a terrace are party walls. There is one between semi-detached houses. Under the *Party Wall Etc Act 1997* 'Awards' are made if work is to be done on a party wall, anything from repairing it to raising it a storey. All this though may be best wrapped up in a retrofit club constitution which facilitates any work necessary on party walls.

Party Wall Surveyors get together in the Pyramus and Thisbe Society!

APPENDIX 12

A national building agency

Not for the first time the reader must indulge the author. Some may now see him mounting his hobby horse.

Are there good ways of building a house? More pertinent to this quest, are there good ways of retrofitting a house to get it nearer to net zero carbon? Whilst keeping options open, are there lessons that might be both learned and shared? And are there, say roof features, or in other elements that should best be avoided ?

Or are we in new territory and in need of sharing experience thinking there is not enough time to let new good practice just permeate out?

Well, the author used to be a structural engineer at The National Building Agency (NBA). This was set up by Geoffrey Ripon secretary of state for housing and local government in 1963. It was closed by Michael Heseltine secretary of state for the environment in 1981.

A main part of its first brief related to new systems of residential building where compliance with building regulations, which tended to relate to brick construction, could hardly apply. So it granted licences for many medium rise building systems.

With high rise it seems to have been expected that capable consulting engineers would design the building and that there would be adequate supervision on site. In the event the latter was not necessarily the case and following a domestic gas explosion parts of the 18 storey Ronan Point in east London collapsed. Jointing between prefabricated concrete panels was found to be inadequate to protect against 'disproportionate collapse'. A nation-wide programme of remedial strapping to similar buildings was put in place. And it was quietly done, though not necessarily straightforward.

The National Building Agency, which was multidisciplinary with architects, structural engineers, surveyors and even economists went on to other ways of easing house building into new ways. One instance was metrication of the house building industry, then adopted by the rest of the construction industry. Another was bringing in, together with the Timber Research and Development Association (TRADA), a coherent form of timber frame construction for houses. (Here it should be emphasized that with this type of house no timber or at least no structural timber is exposed externally. They in fact have brickwork outer leaves with special ties onto the timber frame. Many people may think they have a brick house whereas it is a timber frame house clad in brickwork!)

Later the NBA provided services to local authorities too small to have the in-house expertise that the Greater London Council and other large cities had.

Once central government effectively decreed that local authorities no longer had to build houses and were to help residents with the right to buy them, the NBA had less of a future. HMG asserted that the private market provided similar services. But there were few or no such multi-disciplinary practices, only one or two that were NBA offshoots.

Don't we need a National Building Agency or a National Retrofit Agency now to guide us rapidly towards best retrofit practice? There surely isn't time for this to emerge in 'vernacular' ways which market forces follow?

See also the chapter on previously altered housing.

APPENDIX 12

Decarbonising other UK buildings

This section attempts to provide some background for buildings other than English and Welsh houses.

Bungalows, flats and Scottish dwellings: Some of the presentations in this book will apply, some won't.

Other buildings which are not dwellings and not in the public sector include shops, office buildings, factories, private schools, railway stations, churches, church buildings, zoos and port and airport buildings. Some of the presentations in this book may at times apply, much of it will not though decarbonizing will still be needed.

On 5th November 2022, *Sunday with Laura Kuenssberg* on BBC tv reported that the cost of decarbonising UK public sector buildings (getting them to net zero) is estimated to be £25-£30bn, using government figures. It is not entirely clear whether this includes for instance schools. Possibly hospitals and prisons are not part of this bag either.

The amount was revealed following a Freedom of Information request by the *Sunday with Laura Kuenssberg* show.

The government said the "indicative" figure is based on today's prices and should not be seen as the actual budget needed to move to low carbon heating. Transitioning from fossil fuel heating systems is one way the UK can meet its aim of net zero carbon emissions by 2050. The government has set a target of reducing greenhouse gas emissions from public buildings by 75% by 2037 as part of its net zero strategy.

Schools are a special issue since they are often old and sometimes decrepit. Recent reports (early 2023) are that the government has a long list of schools that are in urgent need of repair, some of which may be at risk of partial collapse. Around 1980 the roof of a school on Stepney Way in London collapsed. Fortunately, this was early in the morning before anyone had arrived. Other roofs of this type remain in use. People might hope that school buildings will be de-carbonised. People might also expect school buildings to be safe! A prime-minister may though have the gall to say safety depends on the economy! If so maybe they will not be re-elected…

APPENDIX 14

THE GRAND DRAMA OF RETROFIT and the 'Cast List'

In some ways retrofitting houses on your street, almost all of them, will be a grand drama. For its duration new ways of making everyone's lives function will have to be got into place. Firstly, road traffic will have to be tightly controlled and in many cases normal parking of cars will be impossible, hence the setting up of a park'n'ride scheme. Clear safe pedestrian routes will be essential. There may have to be gates a bit like a level crossing on a railway line in the country so that heavy plant is got into position and pedestrians are not hurt. Indeed, these sort of manoeuvres may best take place once children are at school.

To illustrate the many conflicts - and opportunities - that might be possible a 'cast list' of the players in your local mass retrofit drama follows.

There may well be differences between points of view and the approaches to take. But decisions will have to be taken, making allowances for differences. Do not be surprised that one cast member is Group Therapist!

Above all kindness will be the watchword. It may do no harm to think that all this is to save our beloved Planet Earth, so almost an act of faith.

CAST LIST (including non-person members!)

1. Resident house owners, 50 or more for any particular mass retrofit,
2. House residents, who are not house owners but are their children or relatives,
3. The sick and frail,
4. Those dying;,
5. Babies,
6. Those about to have a baby,
7. Nurses,
8. Nurses who plan ahead, so the builder's rubble does not get near the blankets.
9. A GP and another GP.
10. Building Societies,
11. Retrofit Building Socuieties,
12. Finance houses; big ones who pay for all this having set up 15 year or so mortgage type plans. Following retrofit they recoup with charges equal to the electricity and gas not used.
13. Insurers who oversee the designs used for the scheme
14. Assurers who warranty the changes made to your house,
15. Insurers who are ready for surprises,
16. Actuaries who do useful sums.
17. Re-insurers who are behind the scenes, ready for the unexpected.
18. Fun. To use up the money you have saved.
19. Builders, but more particularly
20. Big builders, especially
21. Retrofit builders, but indispensably
22. Mass Retrofit builders who form a new industry. Mass Retrofit will be an industry that will be bigger than the construction industry is now. It will use skills the building industry is familiar with and new ones. It will

be in and out of your kitchen almost while the kettle boils.
23. Etiquette officers, after all the mass retrofit may for a time be in your bedroom, though most work above ground storey will be external.
24. Architects,
25. Structural and civil engineers,
26. Building surveyors, who are sometimes better than architects and structural engineers. They will need to survey the premises in detail and to scale. Or have artificial intelligence do this. One benefit will be that it is known whether the furniture will fit back in or not!
27. Quantity surveyors: the old name for people who count up how much the builder is to be paid.
28. Building physicists: a new profession who tell you where your dew-points are etc.
29. Party Wall Surveyors.
30. The workers.
31. The craftspeople (formerly craftsmen).
32. Supermarkets: if you trust a supermarket enough to eat the food you buy from them you may trust them to set up a company for retrofitting houses.
33. Liaison Persons who come a month or more in advance and tell you and yours how it is all going to happen, and allow for your idiosyncrasies.
34. Therapists and counsellors, because for some it will be too much.
35. Group therapists.
36. Samaritans, because for some it will, regrettably, be far too much.
37. Cleaners: there will be a lot of cleaning to be done.
38. Authors: someone should write a book on what mass retrofit feels like[31].

---

[31] In Alban Berg's opera *Lulu*, Alwa, a composer, says someone should write an opera on it.

39. Composers: someone should write an opera on mass retrofit.
40. Producers of soap operas: someone should write a soap opera on mass retrofit. Then people whose houses are not so far retrofitted can see the fun!
41. Traffic officers: people who arrange where the traffic goes because containers are parked in the street where the traffic went.
42. Cranes and their drivers: to lift containers and things.
43. Hotels. For a small minority of families, mass retrofit will turn them out of their house at short notice should significant defects are found at a late stage. Ideally this should be only for a few weeks.
44. Hostels: the kids staying at school four nights a week might not be a bad idea (if the school, those who manage it, and the government are ready).
45. Taxi drivers, to take the kids to school.
46. Electric taxi drivers, to set a good example.
47. Pubs, because they are sometimes in terraces to have Mass Retrofit.
48. Corner shops, because they are sometimes in terraces that are to have Mass Retrofit.
49. A vet.
50. Another vet.
51. Cats and dogs homes: your cat might not like Mass Retrofit, your dog might lift its leg. Cats and dogs do not fit into mass retrofit very well at all.
52. Lawyers: if things are not straightforward.
53. The Police: because some whose houses are having Mass Retrofit are criminals.
54. Funeral directors: because people will die during Mass Retrofit (we may have been going to die next year anyway).
55. Gardeners: at least a part of your garden will be a mess following Mass Retrofit. If you have a nice garden a special gardener may help it survive.

56. Wild people: who say No to Mass Retrofit.
57. People who are not wild: and who say No to Mass Retrofit.
58. The vicar, minister or imam.
59. The Secretary of State for Mass Retrofit (because no government has properly addressed the issue and it is to be hoped they soon will).
60. The Chancellor of the Exchequer, whose job is not to charge VAT on mass retrofit and preparations for it.

APPENDIX 15

Letter to *The Guardian*, February 2023

It is truly shocking that a person has been tried while being prevented from explaining in court their motivation to the jury (Insulate Britain activist jailed for eight weeks for contempt of court, 7 February). By what definition is this justice? There are possible defences in law to the charge of causing public nuisance in these circumstances, but David Nixon – who had taken part in a road-blocking protest for Insulate Britain – was unrepresented in court.

It is clear to everyone, and should have been clear to the judge, that climate catastrophe will cause considerably more nuisance to all those in the court, and outside the court, than the actions of a climate activist. Nixon was acting on my behalf; we all need to be protected from a myopic government that is hellbent on growth at any cost.

DL, Kingston upon Thames, London

BIBLIOGRAPHY

- *Common Building Defects*, diagnosis & remedy, Longman on behalf of National Building Agency 1983
- *EnerPHit: A step-by-step guide to low energy retrofit'* James Traynor; RIBA Publishing, 2019.
- *PAS 2035:2019 Retrofitting dwellings for improved energy efficiency – Specifcation and guidance.* British Standards Institution, 2019.
- *Tristram Shandy* by Laurence Sterne (1713 – 1768, to show how the English can tackle absurdity). The hero is not even born until volume 3. Remembrance of old friend Yorick produces two pages of blackness.
- *The Building Regulations for England & Wales.* (Scotland is different. A smaller part of the population live in houses where the retrofit here described is feasible.)
- *The environmental design pocketbook*, by Sofie Pelsmakers. RIBA Publishing, 2015.
- *The Earth Transformed -an untold history*, Bloomsbury Publishing, Peter Frankopan, 2023

GLOSSARY

1920 Everything changed around 1920. Many bricklayers did not come back from World War I. And the national economy took a hit. However, within a few years local authorities set about housebuilding on a large scale: 'homes fit for heroes', these often influenced by the garden city movement (out of which also came the National Trust). Enter the 20th Century! Portland cement as a principal component of mortar in which bricks were laid resulted in stronger (though more brittle) brickwork. The big builder was no longer unusual.

Cold bridge: A cold bridge is a place where the cold gets in noticeably, really where heat leaks out severely. Some practitioners have thermal imaging cameras that show these locations. But don't expect them to say how to properly get rid of the cold bridge. Indeed, in some buildings there may well be locations where some cold bridges can hardly be avoided, at least for the present. Should a bigger budget become available in future or new technology developed it may be possible to 'rectify' these. An instance could be lengths of rainwater down pipes which are 'embedded' in cladding.

Elevation: the face of an external wall, from ground level to the roof.

EnerPHit: stage by stage conversion of an existing building to get its use to zero-carbon. Trying to achieve Passivhaus effects in an existing building.

Joist: a timber beam in a system of parallel ones that support floorboards.

Party wall: a wall separating two houses.

Passivhaus. This is a term originating in Germany for a newly built house so designed that in both building and use it adds no carbon to the atmosphere, or net-zero carbon to use the jargon.

Purlin a roof timber which supports rafters near their mid-point. In line with the king's regulations following the great fire in London of 1665 they generally were not built into the party walls. Instead, they may be on brick brackets or corbels, or these stout timbers were supported by two or three or more raking (inclined) struts.

Rafters, timbers, often 5" x 2" at 14" centres, extended from the front or rear walls up to the *ridge*. Rafters acted as a double span beam, supported half-way up by a purlin.

Spine wall: a loadbearing wall approximately half way between front elevation and rear elevation of, usually, a pre 1920 house.

Storey: the portion or height of a building between one floor surface and the next. Nowadays, thanks to the National Building Agency, this is most likel 2.62metres. In the later Victorian period rooms were taller since there had to be a minimum distance above a gas light and a six foot person was meant to keep clear of these!

Struts: Struts, literally a stick supporting something, were housed around the purlin halfway up the roof at upper ends and, met the other side purlin on top of the *spine wall* at lower ends. In this way much of the roof weight and occasionally as much again of snow load were placed on to the spine wall.

ILLUSTRATIONS

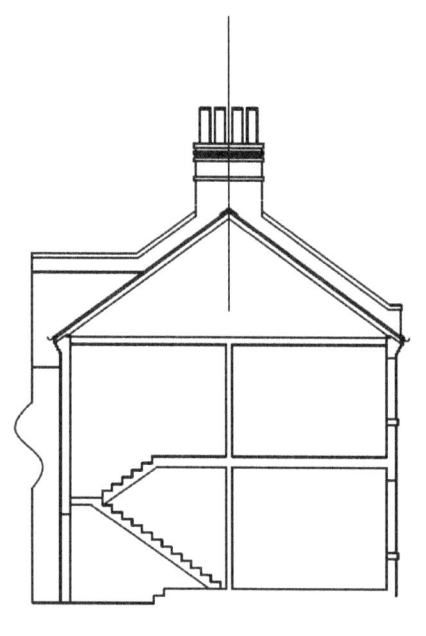

CROSS SECTION THROUGH HOUSE

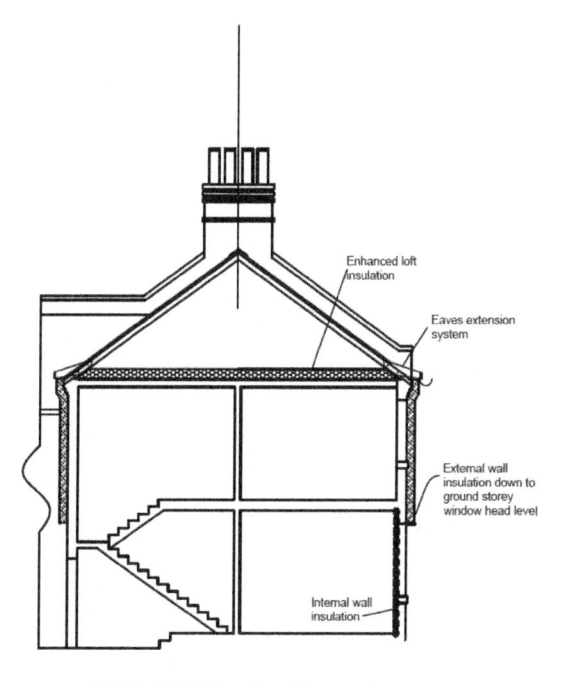

CROSS SECTION INSULATED HOUSE

INDEX

1920, **49, 51, 52, 68, 75, 81, 120, 123, 129, 161, 162**
2028, **19**
2040, **19, 107, 108**
22nd Century, **40, 77, 80**
adjoining house, **24**
aerogel, **65, 82, 90**
alter, **23**
alteration, **23**
Anxieties, **9, 21**
architect, **23, 30, 120, 123, 141**
Artificial intelligence', **38**
asbestos, **68, 125, 126**
Bank of England, **18**
bay window, **23, 137**
boiler, **14, 90**
bracing, **72, 73, 74**
bricklayers, **52, 161**
Bristol, **124**
broker, **9, 21, 24, 26**
budget, **23, 31, 161**
building costs, **34**
building regulations, **23, 145**
Building Regulations, **37, 70, 159**
building society, **26, 34, 45**
Building Surveyors, **31**
bungalows, **22, 72, 113**
butterfly roof, **71**
carbon emissions, **17**
cavity wall, **52, 53**
ceiling joists, **22, 28, 29, 67, 69, 79, 96**
cement, **17**
cement render, **78**

central government, **40, 57, 72, 121, 146**
central valley roof, **55, 71**
central valley roofs, **70**
chartered surveyors, **30**
chimney breasts, **10, 43, 51, 74, 75, 76, 77, 127, 135**
chimney stack, **74**
civil engineering, **22**
cladding systems, **53**
clay tiling, **68**
cold bridge, **23, 49, 56**
complexities, **24**
condensation, **9, 19, 43, 63, 107, 114**
conserve through ventilation, **80**
constitution, **33, 142**
consultant, **9, 24, 26, 30, 104**
cost, **21, 24, 26, 31, 33, 47, 48, 77, 91, 149**
cut home emissions, **15**
dew point, **57, 63**
dew-points, **22**
Do No Harm, **63**
downpipes, **55, 62, 63, 93**
drafts, **10**
draughts, **18, 87, 92**
drone camera, **38, 53**
dying, **26, 35, 152**
east London, **1, 22, 64, 146**
eaves, **23, 56, 57, 93**
energy, **13, 17, 18, 19, 26, 35, 64, 78, 92, 107, 114, 115, 159**
Energy Companies Obligation, **35**

energy usage, 104
EnerPHit, 18, 159, 161
Enhanced Retrofit Passports, 37, 104
everyday life, 25, 26
excess humidity, 80
existing houses, 18, 51
External wall cladding, 53
eye-in-the-sky, 38, 48, 97
Faraday Cage, 129
fees, 31, 33, 34
First Mass Retrofit, 30
flat roof, 22, 136
flats, 22, 113, 115, 119, 120, 149
fossil fuels, 17
funding, 9, 26, 27, 30, 33, 35, 45, 84
funeral, 89, 98
galvanising, 72
general meeting, 9, 21, 22, 30, 34
Great Fire of London, 63
Grenfell, 22, 37, 54
Grenfell Tower, 22, 37, 54
ground floors, 10, 51, 81
gutter extension, 56
Health & Safety Executive, 42
heat pumps, 14
HeLO, 87
HeLOs, 26, 27, 102
highly shrinkable clay, 42
house gutters, 55
House of Commons Committee on Climate Change, 21
house owners, 23, 27, 89, 94, 107, 152
Household liaison officers, 87
hybrid effects, 69

ignorant alteration, 69
ill, 24, 26, 31
initial fee, 23
insulation, 91
Insulation, 28
insurance, 9, 21, 27, 31, 33, 35, 36, 42, 44, 45, 46, 47, 70, 73, 77, 96, 105
insurers, 9, 31, 33, 35, 45, 46, 48, 70, 98, 99, 101, 152
interest, 27, 49
interest rates, 27
internal wall cladding, 10, 87, 90
into the project, 24
JCT, 25
Land Registry, 26, 34, 103
Liaison, 26, 27, 84, 153
licence, 54
lime render, 78
local authorities, 28, 40, 120, 121, 125, 146, 161
local authority building control, 37, 41, 42, 70
loft insulation, 10, 22, 51, 78, 79, 81, 130
London, 22, 28, 31, 40, 46, 55, 63, 64, 66, 70, 71, 76, 84, 88, 98, 119, 121, 141, 142, 146, 162
London Stocks, 64
managing consultant, 21
material change, 70
moisture, 22
Money, 26
mortgage, 26, 34, 45, 96, 152
moth proof, 83
moths, 81
mould, 19, 43, 44, 107, 114

multi-disciplinary, 24
MVHR, 18
neighbour's, 130
net-zero, 104
new town, 22
niceties, 74
old and infirm, 35
opening-up, 73
over cladding, 54
over-cladding, 24, 44, 55, 56, 58, 62, 91, 93, 94
owner-occupiers, 27, 40
owners, 24
ownership, 33
party wall, 43, 71, 73, 75, 77, 96, 123, 135
Party Wall Etc Act 1996, 77
Passivhaus, 17, 18, 23, 28, 93, 102
Passport, 104, 105
Passports, 40, 103
perfection may not be achievable, 62
photogrammetry, 38, 53
pilot phase, 19
pitched roof, 22
planning consent, 41
polystyrene bead fill, 103
Portland cement, 17, 52, 75, 78, 81, 161
professional advice, 23
professionals, 9, 28, 33, 35, 46, 51
purist, 77
Purlin, 66, 162
quilting, 28, 56, 71, 79
Rafter, 66
Rafters, 162
rainwater down pipe, 161

rainwater down pipes, 161
rain-water gutters, 55
reinforcement, 39, 136
repair, 13, 42, 53, 97, 114
replacing all UK houses, 19
retrofit builder, 13, 28, 34, 51, 91, 99, 101
retrofit building society, 13, 34
retrofit club, 9, 21, 24, 26, 28, 33, 43, 46, 49, 84, 85, 93, 101, 102, 142
retrofit contractor, 15, 23
retrofit industry, 15, 54
risks, 9, 26, 27, 33, 45, 69, 92
roof, 10, 14, 22, 28, 29, 41, 51, 55, 56, 66, 67, 68, 69, 70, 71, 72, 73, 74, 75, 79, 80, 91, 96, 97, 124, 130, 137, 139, 145, 161, 162
roofs, 66
search and remediate, 9, 33, 42
semi-detached house, 24
serious defects, 53
Sheffield, 117, 129, 132
snow, 51, 67, 68, 69, 97, 130, 162
snow loading, 67
solar panels, 10, 14, 51, 73, 107
SPAB, 78, 95
specialist stitching, 42
spine wall, 64, 67, 70, 162
steel, 9, 17, 38, 39, 42, 46, 73, 76, 84, 125, 136
structural engineer's design, 70
Strut, 67
subscription, 33, 46
subsidence, 31, 42, 47, 70, 137

survey, 45, 51, 78, 135, 142, 153
Surveyors, 31, 142, 153
temporary roof, 25
tenders, 31, 49
The Event, 10, 15, 87
theodolites, 38
thermal cladding, 13, 53, 83
through-ventilation, 29, 80
toilet facilities, 87, 88
trussed rafter roof, 71
Trussed rafter roofs, 71, 72
umbrella roofs, 90
underpinning, 39, 42
understaffed, 37, 70
U-Value, 53

VAT, 15, 26, 27, 31
warranties, 13, 31, 35, 40, 46
Warranties, 103
warranty assurers, 32
water tank, 10, 29, 51, 79
Welsh slate, 68
WhatsApp, 21
whole street basis, 23
window, 13, 18, 24, 54, 61, 65, 67, 83, 85, 92, 93, 127, 139
wool, 80
Wool, 83
World War, 22, 52, 94, 121, 123, 125, 161
WUFI, 57
zero-carbon, 18, 64, 161

Printed in Great Britain
by Amazon